T0239878

Big Data

Techniques and Technologies
in Geoinformatics

Big Data

Techniques and Technologies
in Geoinformatics

Edited by
Hassan A. Karimi

CRC Press
Taylor & Francis Group
Boca Raton   London   New York

CRC Press is an imprint of the
Taylor & Francis Group, an **Informa** business

CRC Press
Taylor & Francis Group
6000 Broken Sound Parkway NW, Suite 300
Boca Raton, FL 33487-2742

First issued in paperback 2017

© 2014 by Taylor & Francis Group, LLC
CRC Press is an imprint of Taylor & Francis Group, an Informa business

No claim to original U.S. Government works

Version Date: 20140108

ISBN 13: 978-1-4665-8651-2 (hbk)
ISBN 13: 978-1-138-07319-7 (pbk)

Library of Congress Cataloging-in-Publication Data

Big data : techniques and technologies in geoinformatics / editor, Hassan A. Karimi.
    pages cm
  Includes bibliographical references and index.
  ISBN 978-1-4665-8651-2 (hardback)
  1. Geography--Data processing. 2. Big data. 3. Geographic information systems.
4. Geospatial data. 5. High performance computing. I. Karimi, Hassan A. editor of compilation.

G70.2.B54 2014
910.285′57--dc23   2013047353

**Visit the Taylor & Francis Web site at
http://www.taylorandfrancis.com**

**and the CRC Press Web site at
http://www.crcpress.com**

Contents

Preface

What is big data? Due to increased interest in this phenomenon, many recent papers and reports have focused on defining and discussing this subject. A review of these publications would point to a consensus about how big data is perceived and explained. It is widely agreed that big data has three specific characteristics: volume, in terms of large-scale data storage and processing; variety, or the availability of data in different types and formats; and velocity, which refers to the fast rate of new data acquisition. These characteristics are widely referred to as the *three Vs of big data,* and while projects involving datasets that only feature one of these Vs are considered to be *big,* most datasets from such fields as science, engineering, and social media feature all three Vs.

To better understand the recent spurt of interest in big data, I provide here a new and different perspective on it. I argue that the answer to the question of "What is big data?" depends on when the question is asked, what application is involved, and what computing resources are available. In other words, understanding what big data is requires an analysis of time, applications, and resources. In light of this, I categorize the time element into three groups: past (since the introduction of computing several decades ago), near-past (within the last few years), and present (now). One way of looking at the time element is that, in general, big data in the past meant dealing with gigabyte-sized datasets, in the near-past, terabyte-sized datasets, and in the present, petabyte-sized datasets. I also categorize the application element into three groups: scientific (data used for complex modeling, analysis, and simulation), business (data used for business analysis and modeling), and general (data used for general-purpose processing). Finally, I classify the resource element into two groups: advanced computing (specialized computing platforms) and common computing (general-purpose workstations and desktops). It is my hope that analyzing these categories in combination will provide an insight into what big data is, as summarized in the following:

Past-scientific, near-past-scientific, and present-scientific: Big data has routinely challenged scientists and researchers from various fields as the problems are often data intensive in nature and require advanced computing methods, mainly high-performance computing resources (e.g., supercomputers, grids, and parallel computing).

Past-business and near-past-business: While business analysts occasionally had to deal with large datasets, they were faced with limited big data challenges (where the data volume was large and/or fast data processing was required) for which advanced computing, mainly distributed computing (e.g., clusters) and powerful common computing resources, was often utilized.

Present-business: Business analysts are now routinely challenged by big data problems as modern business applications typically call for analysis of massive amounts of data, which might be in various types and formats, and fast analysis of data to produce quick responses. Advanced computing, mainly cloud computing, which is becoming a common computing resource, is used to address these challenges.

Past-general, near-past-general, and present-general: When not addressing science problems or business applications, business analysts are occasionally faced with complex data that overwhelms available resources. In such cases, more powerful common computing resources are considered.

These general considerations are severely exacerbated when big data problems concern geospatial data. This is because geospatial applications are intrinsically complex and involve large datasets; data are collected frequently and rapidly through advanced geospatial data collection technologies that are widely available in mobile devices (e.g., smartphones); and geospatial data are inherently multidimensional. In light of these challenging aspects of geospatial data, this book is focused on big data techniques and technologies in geoinformatics. The chapters of the book, contributed by experts in the field as well as in other domains such as computing and engineering, address technologies (e.g., distributed computing such as clusters, grids, supercomputers, and clouds), techniques (e.g., data mining and machine learning), and applications (in science, in business, and in social media).

Chapter 1 provides an overview of distributed computing, high-performance computing, cluster computing, grid computing, supercomputing, and cloud computing. Chapter 2 describes the Global Earth Observation System of Systems Clearinghouse, an infrastructure that facilitates integration and access to Earth observation data for global communities. Chapter 3 discusses a cloud computing environment (CCE) for processing large 3D spatial datasets and illustrates the application of the CCE using a case study. Chapter 4 describes building open environments as a means of overcoming the challenges of big data in Earth sciences. Chapter 5 discusses the development of visualization and analysis services for NASA's global participation products. Chapter 6 addresses the design of algorithms suitable for geospatial and temporal big data. In Chapter 7, various machine learning techniques for geospatial big data analytics are discussed. Chapter 8 describes the three Vs of geospatial big data and presents a case study for each of them. Big data opportunities in volunteered geographic information to improve routing and navigation services are explored in Chapter 9. Chapter 10 presents a discussion of data mining of taxi trips using road network shortcuts. Big data challenges in social media are outlined in Chapter 11. Chapter 12 presents a pattern detection technique called TCM-Pattern to provide insights into big data. Chapter 13 discusses a geospatial cyberinfrastructure for addressing big data challenges on the World Wide Web. Chapter 14 provides a review of Open Geospatial Consortium (OGC) standards that address geospatial big data.

Editor

Dr. Hassan A. Karimi received his BS and MS in computer science and PhD in geomatics engineering. He is a professor and director of the Geoinformatics Laboratory in the School of Information Sciences at the University of Pittsburgh, Pittsburgh, Pennsylvania. His research is focused on navigation, location-based services, location-aware social networking, geospatial information systems, mobile computing, computational geometry, grid/distributed/parallel computing, and spatial databases and has resulted in more than 150 publications in peer-reviewed journals and conference proceedings, as well as in many workshops and presentations at national and international forums. Dr. Karimi has published the following books: *Advanced Location-Based Technologies and Services* (sole editor), published by Taylor & Francis Group in 2013; *Universal Navigation on Smartphones* (sole author), published by Springer in 2011; *CAD and GIS Integration* (lead editor), published by Taylor & Francis Group in 2010; *Handbook of Research on Geoinformatics* (sole editor), published by IGI in 2009; and *Telegeoinformatics: Location-Based Computing and Services* (lead editor), published by Taylor & Francis Group in 2004.

Contributors

Marc P. Armstrong
Department of Geography
The University of Iowa
Iowa City, Iowa

Jamal Jokar Arsanjani
Institute for Geographic
    Information Science
Ruprecht Karl University of Heidelberg
Heidelberg, Germany

Roland Assam
Data Management and Exploration
    Group
RWTH Aachen University
Aachen, Germany

Mohamed Bakillah
Institute for Geographic
    Information Science
Ruprecht Karl University of Heidelberg
Heidelberg, Germany

and

Department of Geomatics Engineering
University of Calgary
Calgary, Alberta, Canada

Jeffrey Burnett
Department of Computer Science
University of Northern Iowa
Cedar Falls, Iowa

Arie Croitoru
Department of Geography and
    Geoinformation Science
Center for Geospatial Intelligence
George Mason University
Fairfax, Virginia

Andrew Crooks
Department of Computational Science
George Mason University
Fairfax, Virginia

Meixia Deng
Center for Spatial Information Science
    and Systems
George Mason University
Fairfax, Virginia

Liping Di
Center for Spatial Information Science
    and Systems
George Mason University
Fairfax, Virginia

Michael R. Evans
Department of Computer Science
University of Minnesota
Minneapolis, Minnesota

Chih-Yuan Huang
Department of Geomatics Engineering
Schulich School of Engineering
University of Calgary
Calgary, Alberta, Canada

Qunying Huang
Center for Intelligent Spatial Computing
George Mason University
Fairfax, Virginia

Hassan A. Karimi
Geoinformatics Laboratory
School of Information Sciences
University of Pittsburgh
Pittsburgh, Pennsylvania

Steven Kempler
Goddard Earth Sciences Data and
    Information Services Center
Goddard Space Flight Center
National Aeronautics and Space
    Administration
Greenbelt, Maryland

Johannes Lauer
Institute for Geographic
    Information Science
Ruprecht Karl University of Heidelberg
Heidelberg, Germany

Jing Li
Center for Intelligent Spatial Computing
George Mason University
Fairfax, Virginia

Wenwen Li
Center for Intelligent Spatial Computing
George Mason University
Fairfax, Virginia

Zhenlong Li
Center for Intelligent Spatial Computing
George Mason University
Fairfax, Virginia

Steve H.L. Liang
Department of Geomatics Engineering
Schulich School of Engineering
University of Calgary
Calgary, Alberta, Canada

Kai Liu
Center for Intelligent Spatial Computing
George Mason University
Fairfax, Virginia

Zhong Liu
Goddard Earth Sciences Data and
    Information Services Center
Goddard Space Flight Center
National Aeronautics and Space
    Administration
Greenbelt, Maryland

and

Center for Spatial Information Science
    and Systems
George Mason University
Fairfax, Virginia

Lukas Loos
Institute for Geographic
    Information Science
Ruprecht Karl University of Heidelberg
Heidelberg, Germany

Lizhi Miao
Center for Intelligent Spatial Computing
George Mason University
Fairfax, Virginia

Amin Mobasheri
Institute for Geographic
    Information Science
Ruprecht Karl University of Heidelberg
Heidelberg, Germany

Doug Nebert
Federal Geographic Data Committee
Reston, Virginia

Dev Oliver
Department of Computer Science
University of Minnesota
Minneapolis, Minnesota

Dana Ostrenga
Goddard Earth Sciences Data and
    Information Services Center
Goddard Space Flight Center
National Aeronautics and Space
    Administration
Greenbelt, Maryland

and

ADNET Systems, Inc.
Lanham, Maryland

Jacek Radzikowski
Center for Geospatial Intelligence
George Mason University
Fairfax, Virginia

Carl Reed
Open Geospatial Consortium
Wayland, Massachusetts

Thomas Seidl
Data Management and Exploration
    Group
RWTH Aachen University
Aachen, Germany

Monir H. Sharker
Geoinformatics Laboratory
School of Information Sciences
University of Pittsburgh
Pittsburgh, Pennsylvania

Shashi Shekhar
Department of Computer Science
University of Minnesota
Minneapolis, Minnesota

Anthony Stefanidis
Department of Geography and
    Geoinformation Science
Center for Geospatial Intelligence
George Mason University
Fairfax, Virginia

Ramanathan Sugumaran
Department of Geography
The University of Iowa
Iowa City, Iowa

Min Sun
Center for Intelligent Spatial Computing
George Mason University
Fairfax, Virginia

William Teng
Goddard Earth Sciences Data and
    Information Services Center
Goddard Space Flight Center
National Aeronautics and Space
    Administration
Greenbelt, Maryland

and

Wyle Information Systems, Inc.
Mclean, Virginia

Terence van Zyl
Meraka Institute
Council for Scientific and Industrial
    Research
Pretoria, South Africa

Ranga R. Vatsavai
Computational Sciences and
    Engineering Division
Oak Ridge National Laboratory
Oak Ridge, Tennessee

Nicole Wayant
Topographic Engineering Center
U.S. Army Engineer Research and
    Development Center
Alexandria, Virginia

Huayi Wu
Center for Intelligent Spatial Computing
George Mason University
Fairfax, Virginia

Jizhe Xia
Center for Intelligent Spatial Computing
George Mason University
Fairfax, Virginia

Chaowei Yang
Center for Intelligent Spatial Computing
George Mason University
Fairfax, Virginia

Jianting Zhang
Department of Computer Science
City College of New York
New York, New York

Nanyin Zhou
Center for Intelligent Spatial Computing
George Mason University
Fairfax, Virginia

Xun Zhou
Department of Computer Science
University of Minnesota
Minneapolis, Minnesota

Alexander Zipf
Institute for Geographic
    Information Science
Ruprecht Karl University of Heidelberg
Heidelberg, Germany

1 Distributed and Parallel Computing

Monir H. Sharker and Hassan A. Karimi

CONTENTS

1.1   INTRODUCTION

Geoinformatics researchers and practitioners have been facing challenges of geospatial data processing ever since the debut of digital mapping, the predecessor term for today's geospatial information systems. Over time, due to the advancement in data collection technologies (such as satellite imagery and GPS), challenges of geospatial data processing have become widespread. Today, every individual or organization that either routinely or sporadically must handle geospatial data has a better understanding and much appreciation for the complex nature of geospatial data processing. Of several characteristics of geospatial phenomena that make geospatial data to be complex for processing, data-intensive computing, one of the main characteristics of geospatial big data that can be addressed by employing special computing hardware and advanced computing techniques, is discussed in this chapter. To that end, since geospatial big data is defined and different approaches for addressing challenges of this emergent paradigm are discussed in all other chapters of this book, in this chapter high-performance computing (HPC), commonly considered as one main approach for handling data-intensive problems, is focused.

HPC has evolved over time to address the increasing demand of applications for efficient computing. HPC platforms are typically designed for handling problems that are intrinsically data- and/or compute-intensive in nature. Given this trend, HPC is a prominent approach for dealing with some of the big data management and analytics challenges. HPC platforms are basically in the form of distributed and/or parallel computing, which include cluster computing, grid computing, supercomputing, and cloud computing. In sections of this chapter, the basics of distributed and parallel computing and state-of-the-art HPC platforms for addressing geospatial big data problems are discussed.

1.2   DISTRIBUTED COMPUTING

Distributed computing is a reference to computation on a platform with multiple nodes each with its own hardware (e.g., computers) and software (e.g., operating systems) [1]. The nodes in a distributed computing platform could be in close proximity (e.g., in an office building) connected via a local area network (LAN) or dispersed over a large geographic area (e.g., a city) connected via a wide area network (WAN). The computers in a distributed computing platform could be homogeneous, heterogeneous, or a combination of both in terms of operating system, communication protocol, CPU speed, memory size, and storage capacity, among other features. One of the attractive features of distributed computing is scalability, which means that the platform allows participation of a different number of nodes in computation as the demand changes (i.e., it can scale down or up). Increasing the number of nodes in a distributed computing platform is one possible approach for handling large-scale problems. Another feature of distributed computing is that the same service (or job)

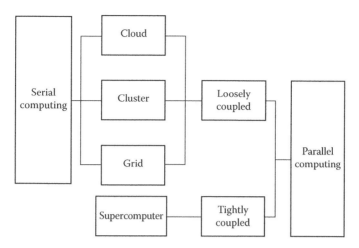

FIGURE 1.1 Computational possibilities in clusters, grids, clouds, and supercomputers.

can be run on multiple nodes simultaneously to increase availability in case of failure of some nodes. The nodes in a distributed computing platform could be tightly coupled or loosely coupled. In the former, the nodes are fixed and removing existing or adding new nodes requires complex procedures, whereas in the latter, the nodes are dynamic and can be removed or added on the fly.

Today, cluster, grid, and cloud computing are commonly distributed computing platforms. Figure 1.1 shows that clusters, grids, and clouds can be utilized for both serial computing and parallel computing, where the nodes are loosely connected. The figure also shows that the architectures of supercomputers are suited for parallel computing, where the nodes are tightly coupled. These distributed computing platforms share similar characteristics in that all perform resource pooling and network access. Cluster and grid computing platforms usually provide services through network access within a private and/or corporate network using LAN/WAN, whereas cloud computing, in most implementations, provides services through public network (e.g., Internet). Each distributed computing platform has its own pros and cons for providing services in terms of domain, cost, and performance and is further described in Sections 1.2.1–1.5.

1.2.1 Cluster Computing

Nodes in a cluster computing platform are connected through dedicated network systems and protocols where they all operate under one centralized operating system [2]. In other words, in a cluster computing platform, interconnected stand-alone computers (nodes) work together as a single computer. The member computers trust each other and share resources such as home directory and computing modules and are equipped with common Message Passing Interface (MPI) [3] implementation to coordinate running a program across computers in the platform. Cluster computing can provide a cost-effective means of gaining desired performance and reliability compared to more expensive proprietary shared memory systems.

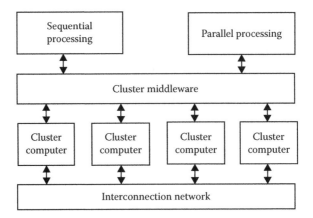

FIGURE 1.2   General cluster architecture.

1.2.1.1 Architecture

A cluster computing platform has at least two computers (nodes) in a single container or separately linked together via a dedicated LAN [2]. In either case, the platform is considered as a single HPC platform to users and applications. The main components of a cluster computing platform are: network, computers, middleware, and user/application. A general architecture of a cluster computing platform is shown in Figure 1.2.

The interconnection network provides the physical connection between the nodes either in one container or through a LAN. Cluster computing platforms are usually designed to be fault tolerant where they can facilitate continuous operation through active nodes. The network interface hardware, connecting the nodes, acts as a communication processor that transmits and receives data packets among the nodes. The interconnection network in a cluster is typically faster than traditional LAN since the communication protocol bypasses the operating system and consequently reduces communication overheads. The nodes can operate collectively as a single virtual machine and/or individually as a computer. The middleware interacts between the nodes and the user programs (serial or parallel) such that the users are given the impression that they are working on a single computer. Programming environments include message passing libraries, debuggers, and profilers providing portable, efficient, and user-friendly tools for application development.

1.2.1.2 Data and Message Communication

The shared memory approach was the means of data and message communication in cluster computing platforms in the 1980s [3]. However, today's cluster computing platforms are based on a new approach called *clustered file system*, which is shared by all the nodes in a cluster computing platform. A clustered file system has redundancy by its nature, but it improves reliability, reduces complexity, and provides location-independent addressing facility [2,4]. However, not all cluster computing platforms take advantage of the clustered file system, as some rely on the physical memory available in each node.

There are two approaches for message communication among the nodes in a cluster: parallel virtual machine (PVM) and MPI [5]. PVM software needs to be installed on each node of the cluster to allow access to software libraries that configure the node as a PVM. For message passing among the nodes, task scheduling, resource management, and fault/failure notification, PVM ensures a run-time environment [5]. MPI, on the other hand, provides various features available in commercial systems in terms of specifications and implementation. MPI is basically a library specification for message passing, proposed as a standard by vendors, developers, and users. MPI implementations typically use TCP/IP and socket communication enabling parallel programming [3].

1.2.1.3 Task Management and Administration

Task management and administration are some of the challenges in cluster computing which at times incur high cost. In particular, task scheduling could be very complex in large multiuser cluster computing platforms when very large datasets need to be accessed. Another challenge is handling heterogeneity of resources on nodes, especially when complex applications are involved. In case of running complex applications on a cluster computing platform, the performance of each task depends on the characteristics of the underlying architecture, how tasks are mapped onto CPU cores, and how the results are integrated. MapReduce and Hadoop are among the latest technologies that can be utilized by cluster computing platforms to improve task management [6].

Handling node failure is one of the common issues in cluster computing where a typical approach, called *fencing* [7], is used to keep the rest of the nodes operational. In other words, upon failure of a node, fencing allows the cluster operate as it was before the node failure. This is accomplished by isolating the failed node in such a way that shared resources in the cluster are not affected by it. Disabling a failed node or denying access from it to shared resources are two possible node isolation approaches [7].

1.2.1.4 Example Geospatial Big Data Project on Cluster

IBM Sequoia is a petascale Blue Gene/Q supercomputer constructed by IBM for the National Nuclear Security Administration as part of the Advanced Simulation and Computing (ASC) Program. Sequoia was delivered to the Lawrence Livermore National Laboratory (LLNL) in 2011 and was fully deployed in June 2012. Sequoia is a 16.72 PFLOPS system running on a cluster platform with 98,000 nodes. Sequoia is planned to be employed for global climate change modeling, which is a geospatial big data project [8] among other big data projects.

1.2.2 GRID COMPUTING

Grid computing is an effective form of HPC that could be a cost-effective and efficient alternative to supercomputers. The main idea behind grid computing is to create a powerful and self-managing virtual computing environment consisting of a large pool of heterogeneous computers that facilitate share of various combinations of resources [9,14]. Grid computing platforms are typically composed of underutilized computing resources, separated geographically but interconnected, which are accessible to users through a single interface providing a single and unified view of the entire resources. The resource-sharing idea is not just about file exchange but direct access to computers, software, data, and other resources. Resource sharing in grid computing is highly protected giving the providers and consumers a clear notion of what can be shared, who is allowed to share, and the sharing conditions. Applications (jobs) can run in parallel (by distribution of tasks on grid nodes) where the results are integrated and sent back to the calling node. In most grid computing platforms, it is possible to allocate all the available nodes to one application for fast processing.

1.2.2.1   Architecture

In a grid computing platform, a user submits job typically from a workstation or a PC to a node known as master. The master node monitors job submissions and could handle as many submissions simultaneously as it is configured for. Upon submission of a job, the master node determines the time and location for the job to be processed by analyzing the requirements of the submitted job and the available resources and their current workloads. The job is scheduled by scheduler, a software available in the master node, that determines which machine should be used to run the job and at what time. The master node then assigns the job to those available nodes that are deemed suitable for processing it. Once the processing is complete in the processing nodes, the master node integrates the results, obtained from the processing nodes, and responds back to the client node.

A typical grid computing architecture has five layers: *fabric, connectivity, resources, collective,* and *application* [10], as shown in Figure 1.3. The fabric layer, which is related to the hardware components, implements local and resource-specific operations and shares the result(s) with the higher layers. The connectivity layer handles communication and authentication protocols. Communication protocols deal with exchange of data among the fabric layer resources, while authentication protocol provides security for both users and resources. The resource layer controls initiations, correspondences, monitoring, accounting, and payment over users and corresponding resources. The fabric layer assists the resource layer with access to and control of local resources. The collective layer contains protocols and services that are global in nature, for example, application programming interfaces (APIs) and software development kits (SDKs), and tracks the interactions among collections

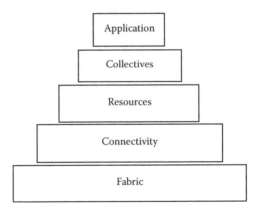

FIGURE 1.3   Grid layer architecture.

and resources. The application layer is where jobs are submitted and the results are delivered/demonstrated. The grid service architecture includes service providers (SPs), service brokers, and service requesters [9].

1.2.2.2 Types of Grid Architectures

Different types of grid architectures are designed to address different types of problems and computing needs. Some grids are designed to take advantage of improved processing resources, whereas some are designed to support collaboration among different resources. Grid architectures can be grouped into three categories according to the functionalities they perform: *information grid, resource grid, and service grid* [9,10].

Information grid collects, organizes, and delivers information of different kinds. For example, global information grid (GIG) is defined as *globally interconnected, end-to-end set of information capabilities for collecting, processing, storing, disseminating, and managing information on demand to warfighters, policy makers, and support personnel* [11]. The information available in information grid is considered always ready to use.

Resource grid coordinates and provides easy, efficient, and transparent access to any resources like computing power, database and/or information base, communication bandwidth, and storage available in the grid irrespective of their locations. The resource grid architecture is further classified into two types: *computational grid* and *data grid* [9,10].

A computational grid aggregates the processing power from the distributed collection of processing nodes to provide a unified processing power of the entire grid computing platform. Computational grids are constructed using cluster of clusters. Computational grids monitor underutilized CPUs, aggregate available computational power, and provide instant access to resources if and whenever required. Other features of computational grids are: automatic deployment, topology resolution, collision resolution, fault tolerance, maintaining checkpoint status, providing grid matrix, pluggability, and data grid integration [10]. SETI@home grid is a well-known example of a computational grid [12]. Computational grids are usually used to compute mathematical equations, determine derivatives, evaluate portfolio, and perform simulation. Scheduling and load balancing are two major tasks that affect performance of computational grids [12].

Data grid provides secure access to distributed heterogeneous pools of data as if user is using a single data source. A data grid makes a group of databases available as a unified single virtual database that is called a federated database [9]. The federated database provides a single-point query service, data modeling, and data consistency. Data grids also connect data, storage, and network resources from distinct domains located in different places; follow local and global policies about data usage; schedule resources efficiently (subject to local and global constraints); and provide high-speed and reliable access to data ensuring data security [9].

Service grid provides services and applications irrespective of their locations, development environment, and hardware specifications. Service grids also correspond with the computational and data grids to provide transparent access to libraries, APIs, SDKs, and applications. Services available on the Internet like search engine, portals, Active Server Pages (ASPs), e-mail, file transfer, and authentication are some of the services that service grid performs [3]. The aforementioned services are supported by service grid where the services may be performed by service nodes in distant geographic locations. People have been sharing their underutilized CPU power through service grid under public-resource computing (PRC) projects [13]. For example, PRC projects, like SETI@home, accumulated 221 million working units in 1 year (starting from July 2001) and the average throughput during that period was 27.36 TFLOPS [3].

1.2.2.3 Topology

Grid architectures are based on different topologies of nodes. Common grid topologies currently available are: *intragrid, extragrid*, and *intergrid* [9]. The simplest of the three topologies is intragrid, which is composed of a basic set of computers (nodes) and services within a single organization. The complexity of the grid design is proportional to the number of entities, within the organization, that the grid is intended to support and the geographical locations and constraints of the participating entities. The more entities join the grid, the more complex are the requirements to ensure security and provide directory services, the less are the availability of resources, and the lower is the performance of the grid. A typical intragrid topology is confined within a single organization, providing a basic set of grid services in the same organization. Here, the nodes are administered and controlled by sharing a common security domain and data internally over a private dedicated network. An intragrid provides a relatively static set of computing resources and can easily share data between grid resources. Intragrids are suitable for organizations that want to handle internal tasks and share data and resources in a secured environment within the organization where there is no interaction necessary with external entity [9].

Extragrid is constructed by connecting two or more intragrids together where typically there are more than one security providers who impose and ensure security according to the organization's policies. Consequently, incorporating these multi-organizational policies result in increased level of management, administration, and security complexities. Extragrid has WAN connectivity where the resources become more dynamic compared to intragrid and the grid needs to be more responsive about node/resource failure. One challenge is workload balancing among the internal and external nodes, especially in case of node failure [9].

Intergrid connects multiple organizations, multiple partners, and multiple clusters; it is basically a grid of grids. It is based on a dynamic integration of applications, resources, and services with partners, customers, and any other authorized organizations that need to obtain access to the grid and its resources through the Internet/WAN. In an intergrid, data are global and public, and applications, both vertical and horizontal, must be developed for a global partner so that anybody can join and be authorized to use the grid [9].

1.2.2.4 Perspectives

A grid could be viewed from different perspectives: user, administrator, and developer [9] as explained in Sections 1.2.2.4.1–1.2.2.4.3.

1.2.2.4.1 User

A new user to a grid computing platform must register to the platform through a specific software available on any of its nodes (computers). In case the user adds a new node to the platform, the user must install the registration software and establish an identity with a certificate authority that has the responsibility of keeping the credentials of the user secure. The user may need to inform the grid administrator about credentials, if any, on other nodes of the grid. From the user perspective, the grid computing platform resembles a large single virtual computer rather than a network of individual computers. The user can query the grid, submit tasks, and check for available resources and monitor progress/status of the submitted task(s).

Tasks submitted to a grid are completed in three phases. First, the input data and the executable program are submitted through one of the grid nodes (client). Sending the input to a grid is called *staging the input data* [1]. Data and programs can also be preinstalled on the grid machines or accessible through a network file system (NFS) that can be mounted on and accessed from any node. Since grid nodes are typically heterogeneous, the grid may register multiple versions of a program so that an appropriate version for each computer could be selected to execute, if and when necessary. Second, the task is submitted and executed on appropriate grid nodes that are monitored to help fault tolerance. Some grids apply a protective *sandbox* for programs to make sure that other components of the grids are protected from any possible malfunctioning of the programs [9]. Third, the results of the job are integrated, if necessary, and sent back to the client. The return of the results could be automatic or through separate mechanisms. If the job requires that a very large number of tasks be performed, the integration of the results might be accomplished over the grid nodes. To avoid large data traffic over the grid, possible options are analyzed beforehand so that unnecessary bottlenecks or deadlock situations do not occur. A networked file system is usually the preferred approach to share data because it facilitates data movement every time the application is run [9].

Users may reserve a set of resources in advance for exclusive or high-priority use. The reservation is granted by taking any future hardware and/or software maintenance schedule for the grid into consideration. At any time, the user may query the grid to monitor the progress of a job and its tasks individually. Along with job scheduling, a grid often provides some degree of recovery for the tasks that fail due to node failure or other reasons. A job may fail for various reasons such as programming error, hardware or power failure, communications interruption, time out, and deadlock. Grid schedulers are often designed to categorize job failures and automatically resubmit jobs so that they are likely to succeed, by running on alternative computers available. In some grids, job recovery could be performed by APIs [9].

1.2.2.4.2 Administrator

Understanding organization's requirements is the prime task of an administrator for planning and choosing an appropriate grid platform. The administrator starts with installing a small grid first and then keeps introducing new components incrementally as necessary. While configuring the grid, the administrator is responsible for ensuring proper bandwidth, recovery plan including appropriate backup, proper arrangement for both private and public key certificates, application software license management, user and donor management, and the highest levels of security in the grid. The administrator is also responsible for managing the resources of the grid including permissions for grid users to use the resources according to the limits as per agreement as well as tracking resource usage and implementing a corresponding accounting or billing. For a large grid, data sharing is another important task of the administrator and all of the resource management issues and concerns apply to data on the grid as well [9].

1.2.2.4.3 Application Developer

Applications can be grouped, from the application developer's perspective, into the following three categories [9] based on the readiness of the applications to run on grid platforms: *ready to use, incapable for grid environment,* and *need to be modified.*

Application developers rely on tools to debug and measure the behavior of grid applications over different grid computers. This is usually accomplished by configuring a small prototype grid so that they can use debuggers on each computer to control and monitor the detailed working behavior of the applications running on grids [9]. Application developers also may use any resources, for example, APIs, offered by the grid.

1.2.2.5 Example Geospatial Big Data Project on Grids

GEON grid (www.geongrid.org) has been developing a 3D model of the Earth's interior. To achieve this, GEON grid collects data, infers interior structures, and evaluates the inferences using computational models. GEON is working to develop a set of software services that can respond to user requests expressed in user's perspectives, such as:

> For a given region (i.e., lat/long extent, plus depth), return a 3D structural model with accompanying physical parameters of density, seismic velocities, geochemistry, and geologic ages, using a cell size of 10 km [11,15].

Integrating a variety of multidimensional data in response to such a request requires a common framework. GEON is developing the OpenEarth Framework (OEF) to facilitate such an integration.

1.2.3 CLOUD COMPUTING

Cloud computing is a nascent HPC paradigm providing hardware infrastructure, computers, software applications, and data as services, all of which available typically as a pay-per-use basis. However, cloud computing is evolved and perceived differently due to its emergent advancement. Gartner [16] states that cloud computing is a computing approach where services are provided over the Internet, which is taking computing to the Internet, using different models and layers of abstraction. Armbrust [17] considers cloud computing as an environment for both the applications delivered as services over the Internet and the hardware and system software in the data centers that provide those services. With these characteristics of cloud computing, both software applications and hardware infrastructures are taken from private environments to third-party data centers and made accessible through the Internet. Buyya [18] considers a cloud as a distributed system comprising a collection of interconnected and virtualized computers. Participating computers are dynamically configured as one or more unified computing resources based on service-level agreements (SLAs). SLAs, which include QoS requirements, are agreements between customers and cloud providers. An SLA specifies the specifications, terms, and conditions of the service to be provided with agreed upon metrics by all parties and penalties for violating the agreement(s).

The US National Institute of Standards and Technology (NIST) defines cloud computing as "a model for enabling ubiquitous, convenient, and on-demand network access to a shared pool of configurable computing resources (e.g., networks, servers, storage, applications, and services) that can be rapidly provisioned and released with minimal management effort or service provider interaction" [19]. NIST identified five essential characteristics, as listed below, in its cloud model [19]:

1. *On-demand self-service*: A user can choose and access computing resources, (e.g., server time, storage, application, data) automatically as needed. There is no need to interact with SP to access resources or scale up/down resource utilization.
2. *Broad network access*: Resources are available over the Internet and can be accessed through standard mechanisms by using heterogeneous client platforms (e.g., PCs, mobile phones, tablets, laptops, and workstations). The client platform can either be thick or thin.
3. *Resource pooling*: SP's computing resources (e.g., storage, processing, memory, and network bandwidth) are pooled to serve multiple users using a multitenant model according to user's demand and are assigned/reassigned dynamically.
4. *Rapid elasticity*: Resources available for users often appear to be unlimited and can be assigned in any quantity at any time as required by users in order to scale up/down the system.
5. *Measured service*: Usage of cloud resources can be monitored, controlled, and reported with transparency for both the provider and user. The cloud platform automatically computes and shares the usage and its cost on a pay-per-use basis.

1.2.3.1 Taxonomies

A four-level taxonomy [20] is presented below to describe the key concepts of cloud computing:

Level 1 (Role): A set of obligations and behaviors viewed by associated actors in cloud

Level 2 (Activity): General behaviors/tasks associated with each specific role

Level 3 (Component): Specific processes, actions, or tasks that must be performed for a specific activity

Level 4 (Sub-component): Modular details of each component

The components in different levels of cloud computing taxonomy [20] are shown in Figure 1.4.

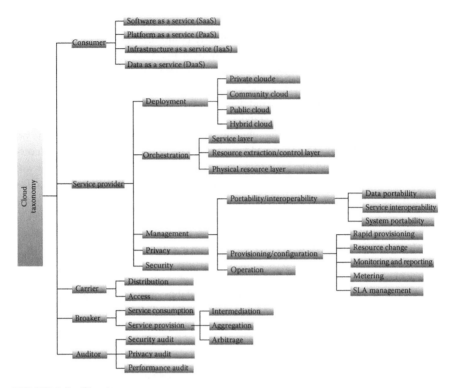

FIGURE 1.4 Cloud taxonomy.

1.2.3.2 Cloud Service Models

Software as a Service (SaaS) provides users with resources to run applications on a cloud platform [19]. User can access applications by using a thin client interface such as a web browser (e.g., web-based e-mail) or by using the interface of the desired application. The user has none or very limited (e.g., application setting) control over the underlying cloud infrastructure (i.e., network, servers, operating systems, storage, or even individual application capabilities) [19]. Examples of SaaS are e-mail, billing, customer relationship management (CRM), and enterprise resource planning (ERP).

Platform as a Service (PaaS) provides users with resources to deploy applications onto the cloud. The applications should be developed by using programming languages and tools supported by the cloud provider [19]. The users are given control over their deployed applications and possibly application hosting environment configurations [19] but not to manage or control the cloud infrastructure. Examples of PaaS are: database development and testing business intelligence, application deployment, and application version control.

Infrastructure as a Service (IaaS) provides users with resources such as processing, storage, networks, and other fundamental computing resources where the user is able to deploy and run both operating systems and application software [19]. However, the users do not have authority over the management or control of the cloud infrastructure but have control over operating systems, storage, deployed applications, and possibly a limited control over selection of networking components (e.g., host firewalls) [19]. Examples of IaaS are: backup and recovery, services management, and storage.

Data as a Service (DaaS) is a specialized service available in some cloud environments where different valuable data are available. Data are collected, stored, updated, provided, and managed by their own data center [20]. However, control over underlying cloud environment is limited to the data unit that is related to the user. Since DaaS is a centrally controlled service model, it is cost-effective, it ensures data consistency, and the option of accessing data in different data structures is available.

1.2.3.3 Cloud Deployment Models

There are four deployments models for cloud: *private cloud, community cloud, public cloud, and hybrid cloud* [19], defined as follows:

1. *Private cloud*: The cloud infrastructure is operated solely for an organization. The infrastructure may be managed by the organization or a third party and it may be located in or outside of the organization.
2. *Community cloud*: The cloud infrastructure is shared by several organizations and supports a specific community that has shared concerns (e.g., mission, security requirements, and policy). It also may be managed by the organizations, a third party, or by both and can be located anywhere.
3. *Public cloud*: The cloud infrastructure is made available to the general public or a large industry group and is owned by an organization offering the cloud services.
4. *Hybrid cloud*: It is a composition of two or more of the aforementioned models (private, community, or public), each of which may be a unique entity, connected such that data and applications are interoperable (e.g., cloud bursting for load balancing between clouds). The interoperability is ensured by a standardized technology to interconnect the participating cloud entities.

1.2.3.3.1 Cloud APIs

One of the attractive features that cloud providers offer for cloud computing solutions is cloud API. APIs provide the primary mechanism for building cloud computing solutions and currently there are five basic categories of cloud APIs that work at four different levels [21].

1.2.3.3.2 Levels of Cloud APIs

Each of the four different levels of cloud APIs [21] requires developer to focus on different tasks and data structures:

Level 1 (the wire): Developers write code directly to the wire format (message, 32 bit words; header, first two words) of the request. For representational state transfer (REST)-based service, the developer creates appropriate

HTTP headers, creates the data payload for the request, and opens an HTTP connection to the service. But for simple object access protocol (SOAP)-based service, the developer creates SOAP envelope instead, adds the appropriate SOAP headers, and fills the body of the SOAP envelope with the data payload.

Level 2 (language-specific toolkits): Developers use a language-specific toolkit to work with SOAP or REST requests at this level. The toolkit helps manage request–response among other details though developers are still focused on the format and structure of the data that is to be sent over the wire.

Level 3 (service-specific toolkits): To work with a specific service, developers use a higher-level toolkit where they are able to focus on business objects and business processes that are far more productive than focusing on the wire protocol, structures, and formats.

Level 4 (service-neutral toolkits): A developer uses a common interface to multiple cloud computing providers. Unlike level 3, developers working at this level are free from complexity of cloud service-specific issues. An application developed using a service-neutral toolkit should be able to use different cloud SPs with minimal changes in the code. The toolkit handles the header and message structure issues and communication protocols.

1.2.3.3.3 Categories of APIs

Five categories of cloud APIs are as follows [21]:

1. *Ordinary programming*: In this category no APIs are specifically developed for any particular cloud. APIs are typically coded in C#, PHP, and Java. They may be used in any cloud environment that supports the language.
2. *Deployment*: APIs that deploy applications to the cloud using cloud-specific packaging technologies such as .Net assemblies.
3. *Cloud services*: Cloud services can be either service specific or service neutral. These APIs are divided into subcategories for cloud storage services, cloud database services, cloud message queues, and other cloud services. In each subcategory, APIs are dedicated only to assigned task.
4. *Image and infrastructure management*: APIs for images are used for uploading, deploying, starting, stopping, restarting, and deleting images, whereas infrastructure management APIs are used to control details, for example, firewalls, node management, network management, and load-balancing tasks.
5. *Internal interfaces*: These are the APIs used to change vendors for the storage layer in a cloud architecture using internal interfaces between the different parts of a cloud infrastructure.

1.2.3.4 Example Geospatial Big Data Project on Clouds

Google manages Google Maps application project on its own cloud [22]. Google Big query, which is a commercial cloud service from Google, also deploys the query for its map project which by itself is a geospatial big data project.

1.3 PARALLEL COMPUTING

The traditional approach for writing computer programs has been monolithic. That is, the software has been written for *serial* computation, one instruction at a time. To run a serial program on a single computer having a single CPU, the program is decomposed into as many single instructions as necessary and then executed one instruction at a time. But in parallel computing, multiple computers are used simultaneously to process submitted jobs. To run in parallel, the job is decomposed into tasks in such a way that each task can be run independently and concurrently on different processors. A parallel computing platform could be a single computer with multiple processors, multiple computers connected by a network, or a combination of both [23]. Accordingly, a program must be *parallelizable* (decomposable into independent tasks) and be able to execute in parallel. That is, the job should be dividable into discrete pieces of independent tasks that can be solved simultaneously. The decomposed tasks should be executed independently in multiple processors and be solved in less time than a single computer would require. Major advantages of parallel computing over serial computing are efficient computation and solving complex problems.

1.3.1 CLASSES OF PARALLEL COMPUTING

One of the most widely used classifications of parallel computing is called Flynn's taxonomy [24]. Flynn's taxonomy distinguishes multiprocessor computer architectures according to two independent dimensions: *Instruction* and *Data*. Each of these dimensions can have only one of two possible states: *Single* or *Multiple*. Flynn's classification [24] is illustrated in Figure 1.5.

Single Instruction, Single Data (SISD): It is a serial computer where only one instruction can run at a time during any one clock cycle of the CPU. Also, only one data stream is being used as input during any one clock cycle. This is a

		Data	
		Single	Multiple
Instruction	Single	SISD single instruction, single data	SIMD single instruction, multiple data
	Multiple	MISD multiple instructions, single data	MIMD multiple instructions, multiple data

FIGURE 1.5 Flynn's classification of parallel computing.

deterministic approach that is the oldest and, even today, the most common type of computer. Examples of SISD are older-generation mainframes, minicomputers and workstations, and most modern PCs such as CRA1, CDC 7600 [24].

Single Instruction, Multiple Data (SIMD): This is a type of parallel computer where all processing nodes execute the same instruction at any given clock cycle and each processing unit can operate on a different data element. SIMD is best suited for specialized problems characterized by a high degree of regularity, such as graphics/image processing. Synchronous (lockstep) and deterministic execution are achieved using two approaches: processor arrays and vector pipelines [24]. Examples of processor arrays are Connection Machine CM-2, MasPar MP-1 and MP-2, and ILLIAC IV and examples of vector pipelines are IBM 9000, Cray X-MP, and NEC SX-2.

Multiple Instructions, Single Data (MISD): This is another type of parallel computer where each processing node operates on the data independently via separate instruction streams and a single data stream is fed into multiple processing units. The only known MISD is the Experimental *C.mmp computer* [35] initiated in Carnegie Mellon University in 1971.

Multiple Instructions, Multiple Data (MIMD): A type of parallel computer where every processor may execute a different instruction stream or work with a different data stream. Execution can be synchronous or asynchronous and deterministic or nondeterministic. MIMD is currently the most common type of parallel computer as most modern supercomputers are based on it. Many MIMD architectures also include SIMD execution subcomponents. Examples of MIMD are most current supercomputers, networked parallel computer clusters and grids, symmetric multiprocessor computers, and multicore PCs [24].

1.3.2 Shared Memory Multiple Processing

Shared memory parallel computers vary widely but generally have the ability for all processors to access all memory as global address space. Multiple processors can operate independently but share the same memory resources. Changes in a memory location effected by one processor are visible to all other processors. Shared memory machines can be divided into two main classes based upon memory access times: *Uniform Memory Access (UMA)* and *Non-Uniform Memory Access (NUMA)* [25].

UMA are common in symmetric multiprocessor (SMP) machines. Processors in UMA are identical and have equal access and access time to memory. UMA is also referred as cache-coherent UMA (CC-UMA), which means if one processor updates a location in shared memory, all the other processors are notified about the update. Cache coherency is accomplished at the hardware level.

NUMA machines are often constructed by physically connecting two or more SMPs where one SMP can directly access memory of another SMP. Unlike UMA, not all processors have equal access time to all memories. Memory access across link takes longer time than memory access from same SMP. In case of cache coherency, it may also be called cache-coherent NUMA (CC-NUMA) [25].

1.3.3   DISTRIBUTED MEMORY MULTIPLE PROCESSING

Like shared memory systems, distributed memory systems vary widely but share a common characteristic, that is, they require a communication network to connect inter-processor memory. Processors in distributed memory systems have their own local memory. Memory addresses in one processor do not map to another processor, so there is no concept of global address space across processors. Because each processor has its own local memory, it operates independently and changes made to a processor's local memory have no effect on the memory of other processors. Hence, the concept of cache coherency does not apply in this case. When a processor needs access to data in another processor, the programmer needs to explicitly define how and when the data are communicated. Tasks synchronization is also the programmer's responsibility. The network *fabric* used for data transfer varies widely, though it can be as simple as the Ethernet [25].

1.3.4   HYBRID DISTRIBUTED SHARED MEMORY

The largest and fastest computers in the world today employ both shared and distributed memory architectures. The shared memory component can be a cache-coherent SMP machine and/or graphics processing units (GPU). The distributed memory component is the networking of multiple SMP/GPU machines, which know only about their own memory, not the memory on another machine [25]. Therefore, network communications are required to move data from one SMP/GPU to another. Considering the current trend, this type of memory architecture will continue to prevail and increase at the high end of computing in the foreseeable future.

1.3.5   EXAMPLE GEOSPATIAL BIG DATA PROJECT ON PARALLEL COMPUTERS

Cosmology@home is a project that seeks to learn how the current universe took its shape [21]. The project manages and executes large numbers of *simulation packages*, each simulating the universe with assumptions about the geometry of space, particle content, and *physics of the beginning*. The simulation results would be compared to real-world observations, in order to see which (if any) of the initial assumptions best matched the real world. In order to produce meaningful results in reasonable time, the project needs to run millions of these simulations in parallel.

1.4   SUPERCOMPUTING

Introduced in 1960s, supercomputers are considered to be among the fastest computer systems at any point of time. In the 1970s, supercomputers featured only a few processors, whereas in the 1990s, they were available in thousands of processors. By the end of the twentieth century, massively parallel supercomputers with tens of thousands of *off-the-shelf* processors were the norm for supercomputers [26,27].

In supercomputing design architecture, systems with a massive number of processors usually are arranged in two ways: *grid* and *cluster* [26]. Also, *multicore* processors in one bundle, called massively parallel processor (MPP) architecture, are another approach of processor arrangement. *Tianhe-2* supercomputer that is developed by China's National University of Defense Technology (NUDT), as of writing this chapter, is the fastest supercomputer in the world that is capable of 33.86 PFLOPS (quadrillions of calculations per second). It will be deployed at the National Supercomputer Center in Guangzhou, China. It took the top position by outperforming *Cray Titan* supercomputer that has a capability of 17.59 PFLOPS [28].

1.4.1 SUPERCOMPUTING WORLDWIDE

As per the latest official record (June 2013), the supercomputing system and performance share worldwide, according to top 500 list, is shown in Figure 1.6. It shows that the United States has 50.4% system share and 47.8% of performance share compared to other countries in the world [29].

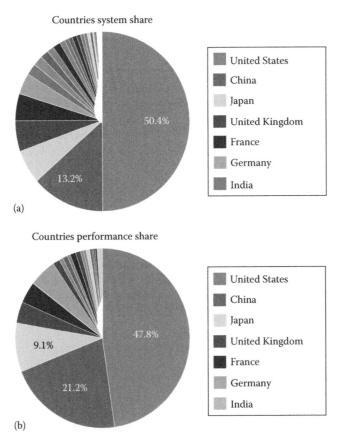

(a)

(b)

FIGURE 1.6 Supercomputing system and performance share worldwide as of June 2013. (a) System share and (b) performance share. (From Top 500 Supercomputer Sites, List statistics, 2013, http://www.top500.org/statistics/list; Buyya, R. et al., *Cloudbus Toolkit for Market-Oriented Cloud Computing*, Cloud Computing, Springer, Berlin, Germany, pp. 24–44, 2009.)

TABLE 1.1

Top 10 Supercomputers in the World

Rank	Name	Manufacturer	Country	Total Cores	Operating System	PFLOPS
1	Tianhe-2 (MilkyWay-2)	NUDT	China	3120000	Kylin Linux	33.86
2	Titan	Cray Inc.	United States	560640	Cray Linux	17.59
3	Sequoia	IBM	United States	1572864	Linux	17.17
4	K Computer	Fujitsu	Japan	705024	Linux	10.51
5	Mira	IBM	United States	786432	Linux	8.59
6	Stampede	Dell	United States	462462	Linux	5.17
7	JUQUEEN	IBM	Germany	458752	Linux	5.01
8	Vulcan	IBM	United States	393216	Linux	4.29
9	SuperMUC	IBM	Germany	147456	Linux	2.9
10	Tianhe-1A	NUDT	China	186368	Linux	2.57

Sources: Top 500 Supercomputer Sites, 2013, http://www.top500.org/lists/2013/06/; Armbrust, M. et al., *Above the Clouds: A Berkeley View of Cloud Computing,* Tech. Rep. UCB/EECS-2009-28, EECS Department, U.C. Berkeley, CA, February 2009.

The latest official top 10 supercomputer specifications all over the world are shown in Table 1.1. In the top spot, Tianhe-2 (MilkyWay-2) pushed DOE/SC/Oak Ridge National Laboratory's Titan out of no. 1 to second place. Titan, a Cray XK7 system, was no. 1 in November 2012 with an impressive 17.59 PFLOPS on the LINPACK benchmark [28]. With 1,572,864 cores, Sequoia (rank 3) was the first system with more than one million cores. But Tianhe-2 introduced 3.12 million cores on its TH-IVB-FEP cluster system [28]. Other supercomputers in the top five systems are Fujitsu's K computer installed at the RIKEN Advanced Institute for Computational Science (AICS) in Kobe, Japan (rank 4), and a Blue Gene/Q system named Mira at Argonne National Laboratory (rank 5) [28].

1.4.2 Trend

On November 7, 2011, Fujitsu announced a 23.2 PFLOPS follow-up system for the K computer, called PRIMEHPC FX10. It uses the same 6D torus interconnect and only one SPARC processor per node. The Chinese government announced on October 31, 2012, that they were building a 100 PFLOPS supercomputer named Tianhe-2, slated for completion in 2015 [30]. A version of Tianhe-2 (MilkyWay-2) with 33.86 PFLOPS was announced on June 17, 2013 [28].

Given the latest speed of progress, industry experts estimate that supercomputers will reach 1 EXAFLOPS (one quintillion FLOPS) by 2018 [30]. At the same time, as their 100 PFLOPS announcement, China also has stated plans to have a 1 EXAFLOPS supercomputer online by 2018. Using the Intel MIC multicore processor architecture, which is Intel's response to GPU systems, SGI plans to achieve a 500-fold increase in performance by 2018, in order to achieve one EXAFLOPS. Samples of MIC chips with 32 cores that combine vector processing units with standard CPU have become available. The Indian government has also stated ambitions for an EXAFLOPS range supercomputer, which they hope to complete by 2017. Erik P. DeBenedictis of Sandia National Laboratories theorizes that a ZETTAFLOPS (one sextillion FLOPS) computer is required to accomplish full weather modeling, which could cover a 2-week time span accurately. Such systems might be built around 2030 [30].

1.4.3 FUTURE SUPERCOMPUTER RESEARCH

With the advancements of hardware sophistication, processor speed, and parallel and distributed computation models, future supercomputer research [31], as mentioned below, has been heading to a more challenging era to address compute-intensive problems.

- *Addressing the memory wall problem*: Increase memory bandwidth
- *Internode interconnection*: New internode interconnects that increase bandwidth, reduce latency, and allow for more performance network topologies
- *Optical computing*: Very active research currently to speed up interconnection network communication
- *Quantum computing*: Computing using quantum mechanical phenomena
- *Parallel languages*: Approaches for both data and control parallelism at the application level using parallel algorithms
- *Algorithms research*: Modified and new algorithms that exploit changing supercomputer hardware characteristics

1.4.4 EXAMPLE GEOSPATIAL BIG DATA PROJECT ON SUPERCOMPUTERS

Earthquake simulation is one of the major geospatial big data projects that is deployed on supercomputers. San Diego Supercomputer Center (SDSC) has created the largest-ever simulation of a magnitude 8.0 (M8) earthquake, primarily along the southern section of the San Andreas Fault. About 25 million people reside in that area, which extends as far south as Yuma, Arizona, and Ensenada, Mexico, and runs up through southern California to as far north as Fresno. SDSC provided the HPC and scientific visualization expertise for the simulation, while the Southern California Earthquake Center (SCEC) at the University of Southern California was the lead coordinator in the project. The scientific details of the earthquake source were handled by researchers at San Diego State University, and Ohio State University was also part of the collaborative effort [32].

1.5   XSEDE: A SINGLE VIRTUAL SYSTEM

The Extreme Science and Engineering Discovery Environment (XSEDE) is the most advanced, powerful, and robust collection of integrated advanced digital resources and services in the world [32]. It is a single virtual system that scientists can use to interactively share computing resources, data, and expertise. Initially, XSEDE supports 16 supercomputers and high-end visualization and data analysis resources across the country. It also includes other specialized digital resources and services to complement these computers. These resources will be expanded throughout the lifetime of the project.

1.5.1   RESOURCES

XSEDE is composed of multiple partner institutions known as SPs, each of which contributes one or more allocable services. Resources include HPC machines, high-throughput computing (HTC) machines, visualization, data storage, and gateway systems. Table 1.2 provides a detailed list of XSEDE computing resources at each of the partner sites with links to detailed specifications for each machine [32].

1.5.2   SERVICES

Scientists, engineers, social scientists, and humanists around the world, many of whom at colleges and universities, use advanced digital resources and services every day. Supercomputers, collections of data, and new tools are critical to the success of those researchers, who use them to make life healthier, safer, and better. XSEDE integrates these resources and services, makes them easier to use, and helps more people use them [33].

Digital services, meanwhile, provide users with seamless integration to NSF's HPC and data resources and facilities. XSEDE's integrated, comprehensive suite of advanced digital services will federate with other high-end facilities and resources, serving as the foundation for a national cyberinfrastructure ecosystem [33]. Common authentication and trust mechanisms, global namespace and file systems, remote job submission and monitoring, and file transfer services are examples of XSEDE's advanced digital services. XSEDE's standards-based architecture allows open development for future digital services and enhancements [33].

XSEDE provides the expertise to ensure that researchers can make the most of the supercomputers and tools [33]. XSEDE helps researchers with the following:

- *Extended collaborative support*: Collaboration with individual research groups or with research communities.
- *Advanced hardware and software architecture*: Individualized user experiences, consistent and enduring software interfaces, improved data management, and transparent integration of complex systems into the overall XSEDE infrastructure.

TABLE 1.2
XSEDE Resources and Partners

Resource Name	Site	Manufacturer/ Platform	Machine Type	Peak Teraflops	Disk Size (Tb)	Availability
Wispy	Purdue U	HP DL140g3	Cluster	0.0	0.0	Production through July 31, 2013
GordonION	SDSC	Appro	Cluster	0.0	4000.0	Production through March 1, 2015
Kraken-XT5	NICS	Cray XT5	MPP	1174.0	2400.0	Production through April 1, 2014
Mason	Indiana U	HP DL580	Cluster	0.0	0.0	Production through March 31, 2014
Lonestar4	TACC	Dell PowerEdge Westmere Linux Cluster	Cluster	302.0	1000.0	Production through January 31, 2014
Keeneland-KIDS	Georgia Tech	HP and NVIDIA KIDS	Cluster	0.0	0.0	Production through August 31, 2014
Steele	Purdue U	Dell 1950 Cluster	Cluster	66.59	130.0	Production through July 31, 2013
Trestles	SDSC	Appro	Cluster	100.0	140.0	Production through June 30, 2014
Quarry	Indiana U	Dell AMD	SMP	0.0	335.0	Production through June 30, 2016
Stampede	UT Austin	Dell Dell PowerEdge C8220 Cluster with Intel Xeon Phi coprocessors	Cluster	6000.0	14336.0	Production
Blacklight	PSC	SGI UV 1000 cc-NUMA	SMP	36.0	150.0	Production through June 30, 2013
Keeneland	Georgia Tech	HP and NVIDIA	Cluster	615.0	0.0	Production through August 31, 2014

Sources: https://www.xsede.org/resources/overview; Xsede, 2013, https://www.xsede.org/

- *User portal*: A web interface to monitor and access XSEDE resources, manage jobs on those resources, report issues, and analyze and visualize results.
- *Coordinated allocations*: NSF's high-end resources and digital services.
- *A powerful and extensible network*: XSEDE SP is connected to a Chicago-based hub at 10 gigabits per second and has a second 10 gigabit per second connection to another national research and education network.
- *Specialized community*: Provided services to serve a particular function and allow for rapid innovation and experimentation.
- *Advanced cybersecurity*: Provides users a feel that XSEDE resources and services are easily accessible but secure.
- *Training and education*: XSEDE-based projects, curriculum development, and more traditional training opportunities.
- *Advanced support*: Especially for novel and innovative projects.
- *Fellowship program*: Brings academia to work closely with advanced user support staff at XSEDE SPs.
- *The technology insertion service*: Allows researchers to recommend technologies for inclusion in the XSEDE infrastructure and enables the XSEDE team to evaluate those technologies and incorporate them.

1.5.3 Example Big Geospatial Data Project on XSEDE

Modeling of storm interaction with XSEDE is a geospatial big data project. The project is focused on real-time weather prediction (especially storm prediction) based on weather data collected since 1970 [33].

1.6 CHOOSING APPROPRIATE COMPUTING ENVIRONMENT

For a given problem, it is critical to choose the appropriate computing platform and resources for a cost-effective solution with a good trade-off with performance. There are several factors that a user should consider for choosing an appropriate computing platform for a given problem.

In choosing a suitable computing environment, some of the factors [34] to consider according to user's requirement are listed below. Not all factors may apply for every user's requirement. Also, all of the factors may not be associated with every computing environment.

- Business level objectives
- Responsibilities of the provider and consumer
- Business continuity and disaster recovery
- System redundancy
- Maintenance
- Location of data
- Seizure of data
- Failure of the provider
- Jurisdiction
- Security
- Data encryption

- Privacy
- Hardware erasure and destruction
- Regulatory compliance
- Transparency
- Certification
- Terminology for key performance indicators
- Monitoring
- Auditability
- Human interaction
- Metrics
- Throughput—Service response time after getting input
- Reliability—Service is availability/unavailability frequency
- Load balancing—Elasticity trigger time
- Durability—Data existence time
- Elasticity—The ability for a given resource to upscale infinitely
- Linearity—System performance while load increases
- Agility—Provider's response time while load scales up and down
- Automation—Amount of human interaction
- Customer service response times—Provider's response time to a user request

Since cloud computing is fairly new, a question may be: What exactly distinguishes this new paradigm from other computing platforms such as cluster computing, grid computing, and supercomputing? Five essential characteristics of cloud computing are compared with those of cluster, grid, and supercomputing as summarized in Table 1.3. All four platforms are distributed and share similar characteristics. The similarities relate to resource pooling and broad network access, two criteria that are fundamental in all distributed computing platforms. Network access to cluster and grid computing platforms usually takes place within a corporate network, while the services of a cloud computing platform are usually accessed through the Internet.

The differences between cloud computing on the one hand and grid and cluster computing, on the other, are attributable to various aspects. Resources in grid and cluster computing are generally pre-reserved, while cloud computing is demand driven, that is, operation of cloud computing platforms is geared to users' actual needs. Another difference concerns the *rapid elasticity* criterion, which forms an integral part of cloud computing but is not normally supported by cluster or grid platforms. Service usage

TABLE 1.3

Comparing Cluster, Grid, Cloud, and Supercomputing

	Cluster	Grid	Cloud	Supercomputing
On-demand self-service	No	No	Yes	No
Broad network access	Yes	Yes	Yes	Yes
Resource pooling	Yes	Yes	Yes	Yes
Rapid elasticity	No	No	Yes	No
Measured service	No	Yes	Yes	Yes

only tends to be accurately measured in grid and cloud computing, whereas the majority of cluster computing platforms simply provision rudimentary metering functions.

Choice of a computing platform may depend on different perspectives. User, system developer, application developer, and provider have different perspectives. Tables 1.4 through 1.7 summarize the comparative analysis among cluster,

TABLE 1.4

Comparing Cluster, Grid, Cloud, and Supercomputing: User Perspective

	Cluster	Grid	Cloud	Supercomputer
On-demand self-service	No	No	Yes	No
Rapid elasticity	No	No	Yes	No
Measured service	No	Yes	Yes	Yes
Maintenance	Yes	No	No	No
Security	Yes	Limited	Poor	Yes
Privacy	Yes	Limited	Poor	Yes
User control	Yes	Limited	No	No
Linearity	Poor	Moderate	SLA	SLA
Agility	Minimum	Moderate	Fast	Immediate
Throughput	Low	Moderate	Higher	Highest
Customer service promptness	No	Poor	High	High
Cost	Small	Moderate	High	High
Performance	Low	Medium	High	Best

TABLE 1.5

Comparing Cluster, Grid, Cloud, and Supercomputing: Developer (System) Perspective

	Cluster	Grid	Cloud	Supercomputer
Hardware size	Small	Medium	High	Medium
Software	Simpler	Simple	Complex	Complex
Implementation	Easier	Easy	Complex	Complex
Service deployment	No	Yes	Yes	Yes
Maintenance	Easier	Easy	Harder	Hard
Handling heterogeneity	No	Limited	Yes	Yes
Ensuring privacy/security	Yes	Limited	Poor	Yes
Fault tolerance	No	Poor	Yes	Yes
Providing scalability	No	Moderate	Yes	Yes
Extending service/capacity	Limited	Limited	Yes	Hard
Development effort	High	Medium	Medium	High
Learning curve	High	Medium	Medium	High
Throughput	Low	Moderate	Higher	Highest
Cost	Small	Moderate	Big	Biggest
Performance	Low	Medium	High	Best

TABLE 1.6

Comparing Cluster, Grid, Cloud, and Supercomputing: Developer (Application) Perspective

	Cluster	Grid	Cloud	Supercomputer
Hardware spec	Small	Medium	High	High
Platform	Fixed	Fixed	Flexible	Fixed
App deployment	No	No	Yes	No
Maintenance	No	No	Yes	No
Ensuring privacy/security	High	Limited	Poor	High
Fault tolerance	Low	Low	High	High
Providing scalability	Poor	Moderate	High	High
Throughput	Low	Moderate	High	High
Development effort	High	Medium	Medium	High
Learning curve	High	Medium	Medium	High
Cost	Low	Moderate	High	High
Performance	Low	Medium	High	High

TABLE 1.7

Comparing Cluster, Grid, Cloud, and Supercomputing: Provider Perspective

	Cluster	Grid	Cloud	Supercomputer
On-demand self-service	No	No	Yes	No
Broad network access	Yes	Yes	Yes	Yes
Resource pooling	Yes	Yes	Yes	Yes
Rapid elasticity	No	No	Yes	No
Measured service	No	Yes	Yes	Yes
Maintenance	Yes	Yes	No	No
Heterogeneity	No	Limited	Yes	Yes
Security	Yes	Limited	Poor	Yes
Fault tolerance	No	Poor	Yes	Yes
Privacy	Yes	Limited	Poor	Yes
Auditability	Yes	Yes	Yes	Yes
User control	Yes	Limited	No	No
Linearity	Poor	Moderate	High	SLA
Agility	Minimum	Moderate	Maximum	Immediate
Throughput	Low	Moderate	Higher	Highest
Learning curve	High	Medium	Medium	High
SLA management	Simpler	Simple	Hard	Hard
Cost	Small	Moderate	Big	Biggest
Performance	Low	Medium	High	Best

grid, cloud, and supercomputing platforms from the perspectives of user, system developer, application developer, and provider, respectively.

Compared to other distributed computing platforms such as grids or clusters, cloud computing provides enterprises significantly more flexibility in terms of performance and cost. They can dispense with IT infrastructures of their own and only have to pay for the resources and services they actually use (*pay-per-use/pay as you go*). The IT infrastructures can be dynamically adapted to changes in business requirements and processes with the help of virtualization technologies and service oriented, distributed software systems.

1.7 SUMMARY

Cluster computing differs from cloud computing and grid computing in that a cluster is a group of computers connected by a LAN, whereas clouds and grids are wide scale and often geographically dispersed. In other words, computers in a cluster are tightly coupled, whereas in a grid or a cloud they are loosely coupled. Also, clusters are typically configured of similar computers, whereas clouds and grids are configured of heterogeneous computers.

Supercomputers can be used to solve data- and compute-intensive problems using massively parallel processor architectures. Supercomputers work along with grid and cluster in some environments. XSEDE provides a platform to use multiple supercomputers, if necessary, shared by its partner supercomputers. However, choosing a computer environment to address a problem at hand is critical in terms of cost and performance. An array of these HPC platforms may help one to achieve in-time solution to both compute-intensive and data-intensive (big data) problems.

REFERENCES

1. Hwang, K., Fox, G.C., and Dongarra, J.J. *Distributed and Cloud Computing: From Parallel Processing to the Internet of Things*. Amsterdam, the Netherlands: Morgan Kaufmann, 2012.
2. Bakery, M. and Buyyaz, R. Cluster computing at a glance. *High Performance Cluster Computing: Architectures and System*. Upper Saddle River, NJ: Prentice-Hall, 1999, pp. 3–47.
3. El-Rewini, H. and Abd-El-Barr, M. Message passing interface (MPI). *Advanced Computer Architecture and Parallel Processing*. Hoboken, NJ: John Wiley, 2005, pp. 205–233.
4. High-performance computing. Enhancing high-performance computing clusters with parallel file systems, 2013. http://www.dell.com/downloads/global/power/ps2q05-20040179-Saify-OE.pdf (Retrieved October 6, 2013).
5. Geist, Al. *PVM: Parallel Virtual Machine: A Users' Guide and Tutorial for Networked Parallel Computing*. Cambridge, MA: the MIT Press, 1994.
6. Dean, J. and Ghemawat, S. MapReduce: Simplified data processing on large clusters. *Communications of the ACM*, 2008, 51(1): 107–113.
7. Erasani, P. et al. Client failure fencing mechanism for fencing network file system data in a host-cluster environment. U.S. Patent No. 7,653,682. January 26, 2010.

8. GlobalPost. IBM's 'Sequoia' supercomputer named world's fastest, 2013. http:// www.globalpost.com/dispatch/news/business/technology/120618/ibms-sequoia-supercomputer-worlds-fastest (Retrieved May 30, 2013).

9. Ferreira, L. et al. *Introduction to Grid Computing with Globus*. San Jose, CA: IBM Corporation, International Technical Support Organization, 2003.

10. Mahajan, S. and Shah, S. *Distributed Computing*. New Delhi, India: Oxford University Press, 2010.

11. GEON, 2013. http://www.geongrid.org/index.php/about/ (Retrieved May 30, 2013).

12. SETI@home. What is SETI@home? 2013. http://setiathome.ssl.berkeley.edu/ (Retrieved October 6, 2013).

13. Anderson, D., Cobb, J., Korpela, E., Lebofsky, M., and Werthimer, D. SETI@home: An experiment in public-resource computing. *Communications of the ACM*, 2002, 45(11): 56–61.

14. Prabhu, C.S.R. Grid and cluster computing. PHI Learning Pvt. Ltd., 2008.

15. GEON. Open earth framework, 2013. http://www.geongrid.org/index.php/projects/oef (Retrieved May 30, 2013).

16. Gartner. Special report cloud computing, 2013. http://www.gartner.com/technology/ research/cloud-computing/report/ (Retrieved May 28, 2013).

17. Armbrust, M. et al. *Above the Clouds: A Berkeley View of Cloud Computing*. Tech. Rep. UCB/EECS-2009-28, EECS Department, U.C. Berkeley, CA, February 2009.

18. Buyya, R., Pandey S., and Vecchiola, C. *Cloudbus Toolkit for Market-Oriented Cloud Computing*. Cloud Computing. Berlin, Germany: Springer, 2009, pp. 24–44.

19. Mell, P. and Grance, T. The NIST definition of cloud computing (draft). NIST Special Publication 800.145: 7, 2011.

20. Rimal, B.P., Choi, E., and Lumb, I. A taxonomy and survey of cloud computing systems. *Fifth International Joint Conference on INC, IMS and IDC, 2009. NCM'09. IEEE*, 2009.

21. Cloud Computing Use Case Discussion Group. Cloud computing use cases white paper, 2013. http://opencloudmanifesto.org/Cloud_Computing_Use_Cases_Whitepaper-4_0. pdf, (Retrieved May 25, 2013).

22. Google maps, 2013. https://maps.google.com/ (Retrieved October 6, 2013).

23. Padua, D., (ed.). *Encyclopedia of Parallel Computing*. Vol. 4. Springer, New York, 2011.

24. Flynn, M. Some computer organizations and their effectiveness. *IEEE Transactions on Computers*, 1972, C-21: 948.

25. Hwang, K. *Advanced Computer Architecture*. Tata McGraw-Hill Education, New York, 2003.

26. Hoffman, A.R. and Traub, J.F. *Supercomputers: Directions in Technology and Applications*. Washington, DC: National Academies, 1990, pp. 35–47.

27. Hill, M.D., Jouppi, N.P., and Sohi, G. *Readings in Computer Architecture*. San Francisco, CA: Morgan Kaufmann, 1999, pp. 40–49.

28. Top 500 Supercomputer Sites, 2013. http://www.top500.org/lists/2013/06/ (Retrieved June 18, 2013).

29. Top 500 Supercomputer Sites. List statistics, 2013. http://www.top500.org/statistics/list/ (Retrieved June 18, 2013).

30. Top 500 Supercomputer Sites, 2013. http://www.top500.org (Retrieved June 18, 2013).

31. Graham, S.L., Snir, M., and Patterson, C.A. (eds.). *Getting up to Speed: The Future of Supercomputing*. National Academies Press, Washington, DC, 2005.

32. News Center. UC San Diego team achieves petaflop-level earthquake simulations on GPU-powered supercomputers, 2013. http://ucsdnews.ucsd.edu/pressrelease/uc_san_ diego_team_achieves_petaflop_level_earthquake_simulations_on_gpu_pow (Retrieved May 30, 2013).

33. Xsede, 2013. https://www.xsede.org/ (Retrieved May 28, 2013).
34. Kosar, T. (ed.). *Data Intensive Distributed Computing: Challenges and Solutions for Large-Scale Information Management.* Hershey, PA: Information Science Reference, 2012.
35. Wulf, W.A. and Bell, C.G. C.mmp: A multi-mini-processor. *Proceedings of the December 5–7, 1972, Fall Joint Computer Conference, Part II.* ACM, New York, 1972.

2 GEOSS Clearinghouse
Integrating Geospatial Resources to Support the Global Earth Observation System of Systems

*Chaowei Yang, Kai Liu, Zhenlong Li, Wenwen Li,
Huayi Wu, Jizhe Xia, Qunying Huang,
Jing Li, Min Sun, Lizhi Miao, Nanyin Zhou,
and Doug Nebert*

CONTENTS

2.1 INTRODUCTION

To better understand, record, and predict phenomena related to human lives on Earth, scientists started to collect data about the Earth's surface several hundred years ago (Jacobson et al. 2000). For example, our ancestors recorded the Earth's surface temperature (Kusuda and Achenbach 1965), volcanic eruptions (Robock 2000), hurricanes (Goldenberg et al. 2001), and other parameters of physical phenomena. In the past century, the invention of digital computers for information processing, the launch of satellites, and the evolution of remote sensing technologies have greatly improved our capabilities to observe and record various physical parameters of the Earth's surface (Christian 2005) at a speed of terabytes to petabytes per day.

These big observational data are of great value to different domains within the Earth sciences because they provide baseline data about the Earth at specific time snapshots. They can be used for different applications and scientific studies, such as for constructing a Digital Earth (Yang et al. 2008), responding to emergencies, and aiding policy making (Yang et al. 2010a). Recognizing the demand for these observed datasets and a need for their wider dissemination, the intergovernmental Group on Earth Observation (GEO) was established to build a Global Earth Observation System of Systems* (GEOSS) to help manage, integrate, access, and share global EO datasets to address worldwide and regional problems. Different committees, such as public health, were established to build such a system of systems to benefit different countries by facilitating share and integration of EO data and knowledge (GEO 2005; Christian 2008; Zhang et al. 2010a).

This information explosion and increasing usage demand pose technical challenges for better management and accessibility of the data and information (Yang et al. 2010a). The scope of GEOSS is broad yet it focuses on sharing Earth science resources with users from global communities (Craglia et al. 2008). As a first step, GEOSS identifies problems, solutions, and pilot applications (Lautenbacher 2006) through the development of the GEOSS Common Infrastructure (GCI; GEO 2009) (see Figure 2.1), which serves as a framework to help the integration of GEOSS, discovery facilitation, and assessment and utilization of EO data, services, and models.

Figure 2.1 shows the technical view of the GCI, which links three components: resource repository, resources, and applications. The resource repository is the GEOSS clearinghouse (CLH) where all metadata of resources are maintained; the resources are EO data, models, or services that are distributed over the Internet; applications

---

* http://www.Earthobservations.org/

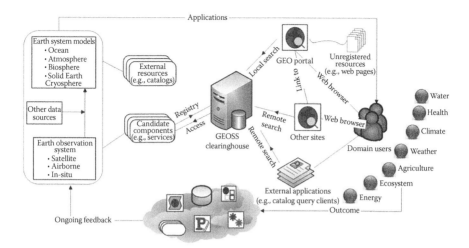

FIGURE 2.1 GEOSS common infrastructure. (Revised from Duchossois, G., The global earth observation system of systems (GEOSS): An example of international cooperation, in 2nd ESA/ASI *Workshop on International Cooperation for Sustainable Space Exploration, Sarteano, Siena,* Italy, May 9–12, 2006, pp. 1–29.)

are systems, tools, or modules deployed for domain users, while the data and service components are online resources that can be discovered from the repository. Interoperability should be achieved between the CLH and the resources and between the CLH and applications. This architecture includes the life cycle of EO data from acquisition, management, application, and decision supports to feedback from the application and decision-making users.

Facing the technical challenges for building such a CLH, GEO organized a CLH competition among different institutions to demonstrate their best solutions to address the challenges. As one of the candidates, the joint center for intelligent spatial computing (CISC) developed a CLH based on an open-source software GeoNetwork (Ticheler and Hielkema 2007; Ticheler 2008; OSGeo 2009) and was selected as the operational CLH* to support GEOSS (GEO 2010).

This chapter reports our research and development of the CLH including literature review (Section 2.2), the challenges (Section 2.3), detailed implementations and functions (Section 2.4), and the status (Section 2.5). Section 2.6 summarizes this chapter and discusses future research.

2.2 CATALOG AND CLEARINGHOUSE RESEARCH REVIEW

2.2.1 METADATA REPOSITORY AND STANDARDIZED METADATA

To better facilitate the sharing of vast amount of domain datasets in the data discovery, access, and visualization process, a variety of research projects have been conducted to develop metadata for describing the datasets and online linkage to access data

---

* http://clearinghouse.cisc.gmu.edu/geonetwork

(e.g., Huang et al. 2011; Li et al. 2013). In this process, metadata, the data content, and online access are standardized through, for instance, Federal Geographic Data Committee (FGDC), International Standardization Organization (ISO 2007) metadata standards (Nativi and Bigagli 2009), and Universal Description, Discovery and Integration (UDDI). These metadata are archived in catalogs to provide searching capabilities for domain users to discover datasets needed (Huang et al. 2011), addressing the data sharing problem in different domains.

2.2.2 CATALOG AND CLEARINGHOUSE BASED ON SERVICE-ORIENTED ARCHITECTURE AND STANDARD SERVICES

To share metadata repository across different systems including catalogs and CLHs, the interfaces of metadata should be standardized as, for example, Catalog Service for the Web (CSW) and Z39.50. The standardized service interfaces are used in a loosely coupled service-oriented architecture (SOA) fashion for sharing among different catalogs (Paul and Ghosh 2008; Yang et al. 2008). Interoperability was achieved through the interface standardization and SOA for supporting a domain application (Bai and Di 2011).

2.2.3 SEMANTIC-BASED METADATA SHARING AND DATA DISCOVERY

Metadata are normally captured by specific domain users for specific usage and shared within a standardized format to achieve interoperability for wider usage. When they are shared across domains or usages, it becomes difficult to discover the right data because of the mismatch between end users and the metadata content (Li et al. 2008). Ontology and semantics are used to capture the meaning of user query and the potential usage of data in a knowledge management and utilization fashion to increase the potential searching results and to refine final results (Zhang et al. 2010b). With specific quality controls, searching results can be further enhanced (Khaled et al. 2010) for chaining services (Yue et al. 2007).

These research projects provide insightful results to help the development of the CLH. CLH bridges resources and users through two major tasks (1) serving as a repository of big metadata covering metadata services from various domains and (2) promoting the use of registered services for different applications. These tasks pose a variety of technological challenges that were not adequately addressed before (Oppen and Dalal 1983; Nebert 2004; GEO 2009; Yang et al. 2010b) and include (1) wide distribution of catalogs, (2) heterogeneity in the metadata, (3) a huge amount of metadata for the vast amounts of EO data collected, (4) frequent updating of the datasets and metadata, (5) duplication of the same or slightly different metadata for same or similar data and services, (6) mismatch between the current metadata category and the categories defined by GEOSS, and (7) acceptable performance when providing cataloging and searching against millions of metadata records to concurrent users. Section 2.3 reports our solution to the challenges, and the solution is further detailed in Sections 2.4 and 2.5.

2.3 TECHNOLOGICAL ISSUES AND SOLUTIONS

2.3.1 INTEROPERABILITY

Interoperability has been a long-standing problem within the geospatial community since the early 1990s when the Internet started to provide convenient channel to share geospatial data (Bishr 1998). Due to the structural and semantic heterogeneity of geospatial data, providing a uniform architecture and interface to share, exchange, and interoperate distributed geospatial data is still a pervasive problem today (Yang et al. 2008; Li et al. 2010). To solve this problem, the Open Geospatial Consortium (OGC) collaborated with other organizations, such as the FGDC, GEOSS, and ISO, to advance interoperability by developing a series of open, community-consensus standards for metadata harvesting and user access (Khalsa et al. 2009). Utilizing the standards, Figure 2.2 demonstrates the interoperability enablement of CLH at the data, service, and catalog level.

Remote geospatial data are encapsulated and deployed to be serviced within various OGC services, shown in the dark gray node as OGC *WxS*. Other raw geospatial data are standardized with the ISO, Organization for the Advancement of Structured Information Standards (OASIS), FGDC, and Dublin Core Metadata Initiative (DCMI) (2010) to promote the discovery and reuse of various geospatial resources (gray nodes within the left column). The metadata of the distributed data resources are registered into the CLH as a data and service container, which provide multiple ways for remote resource discovery by exploiting the consistency of the metadata. CLH supports a series of advanced models and protocols for a cross-system search. These models include CSW (Nebert and Whiteside 2005) and Search/Retrieval via URL (SRU) as the pink nodes in the *Search Standards* column of Figure 2.2. CSW is widely adopted by the geospatial community for metadata-based resource discovery and developing various application profiles, including ISO 19115/19139 and OASIS ebRIM (ebXML Registry Information Model). For cross harvesting multiple CSW catalogs to the CLH, it is required to automatically interpret the structures of remote catalogs, as well as to store the obtained metadata into the backend repository.

In our implementation, the current CLH repository supports the storage of three standard metadata formats: Dublin Core (DC), FGDC, and ISO 19139. When the CLH extracts metadata using the formats, the metadata are directly inserted into the metadata repository. While the metadata are encoded using ebXML, an EXtensible Stylesheet Language Transformation (XSLT) file that maps the corresponding code from current metadata format to the desired ISO 19139 will be loaded. The transformation is performed on the fly before inserting the new metadata into the CLH.

Another information retrieval mode supported by CLH is Z39.50 (NISO 2010), which is the essential component for a library-based catalog. When searching a system, end users provide a standardized query with the complexity and difference of model implementations hidden. After the search message is received by the CLH, it is translated into the syntax that the backend repository understands. Similarly, SRU-based discovery is based on Z39.50 but provides a more flexible response mode. For example, it is XML-based and supports HTTP GET, HTTP POST, and web service features such as Simple Object Access Protocol (SOAP) for message transfers

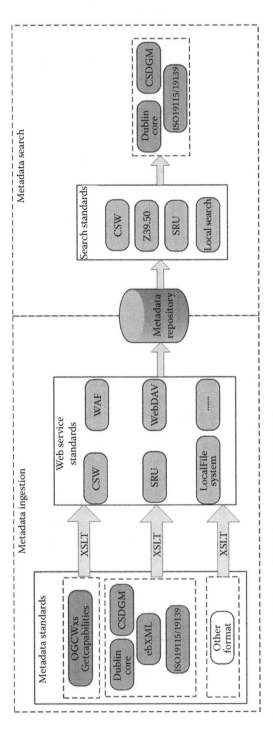

FIGURE 2.2 Interoperability and inclusiveness of GEOSS CLH.

(Zhang and Li 2005). With the flexibility of information retrieval modes provided, the CLH is capable of communicating with other existing catalogs for geospatial data sharing and interoperability.

2.3.2 PROVENANCE AND UPDATING

Data provenance refers to the description of the origins of data and process conducted before the data arrived in a database (Buneman et al. 2001). Recently, many projects and surveys have examined the role of data provenance so as to support efficient information retrieval, multi-document summarization, question answering, and knowledge discovery (Simmhan et al. 2005). However, it is hard for scientists or the general public to select the most suitable data with a high level of confidence. To address this problem, CLH is designed with a corresponding data structure for storing relevant information for data provenance. Two databases are connected within the repository (Figure 2.3). One is to support active data discovery and the other to store historical data records that have been harvested by the CLH. Once a metadata XML is retrieved, the fields (UUID, create date, modified date, data source, URI of the data source, and metadata schema) are extracted accordingly, of which UUID acts as the key index for each record in the metadata table. The dates (create date and date of last change) are stored as an ISO date in the active database.

The dates of the metadata are in different formats, such as RFC822 (standard for the format of ARPA Internet text messages; Crocker 1982), ISO 8601 (Kuhn 2004), or simple Java Date Object. To avoid duplication because of data format, CLH parses and converts all the date formats into the ISO 8601 format. Once a new metadata file is acquired, a UUID-based query is conducted in the database to check its existence. If no record exists, the new record is directly inserted into

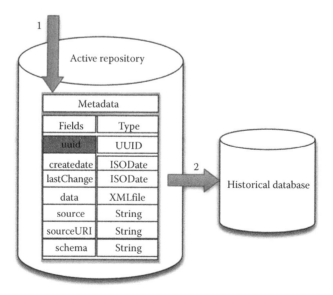

FIGURE 2.3 Updating and provenance of geospatial metadata in GEOSS CLH.

the metadata table; otherwise, the modified dates between the matched records will be compared. When the data records have the same creation date and modification date, no operation is performed. If it is indicated as an updated version, the existed data record will be moved into the historical database and a new record will be inserted into the active database.

With this updating mechanism, CLH can manage data effectively by reducing data redundancy and can provide efficient searches with the support of the provenance information (Li et al. 2010). In addition, there is a significant amount of records similar to the content but with different file identifiers. To avoid duplication, the duplicated records are discarded before ingesting to the CLH.

2.3.3 System Performance

System performance is critical for the CLH operation. If the response time is longer than 3 s, users will become frustrated (Nah 2004). Therefore, maintaining a consistently short response time becomes essential, especially when millions of metadata records are stored in the CLH and the data volume is continually increasing. Indexing is a core solution that has been utilized in existing CLHs, such as the CLH provided by Global Climate Master Directory.* However, the traditional Lucene-based approach opens and closes the index frequently, which leads to a low performance for front-end search especially when batch importing datasets into the CLH.

To address this problem, CLH adopted the following technologies (Yang et al. 2005; Li et al. 2010): (1) buffering, where the index is kept in the RAM when writing a set of metadata rather than writing only one metadata file, and (2) multithreading, where the indexing tasks are scheduled and conducted by multiple threads in parallel. Meanwhile, the modules that operate client search and indexing are encapsulated into separated threads. Once the index is closed for updating, the search thread uses the old index stored in the hard drive; once the update is complete and the new index is written back to the hard drive from RAM, the search thread is released to provide updated search results to the end users. Leveraging these strategies, CLH provides consistently quick responses (1–2 s) for concurrent searches from various clients.

2.3.4 Timely Updating

CLH uses a periodic process for metadata updating, for example, once a day. A periodic time (in days, hours, minutes) between two consecutive harvests from a node are added for the timely updating as illustrated in Figure 2.4. There are active icons showing for the third, fourth, and fifth harvesting nodes. The *Active* status shows that the harvesting engine is waiting for the timeout specified for this node. When the timeout is reached, the harvesting starts. The *Every* shows the time (in days, hours, minutes) between two consecutive harvesting. Together with the change date and the updating mechanism in Section 2.3.2, the periodic harvesting will update the metadata with the most recent version.

---

* http://www.gcmd.nasa.gov/

Select	Name	Type	Status	Errors	Every	Last run	Operation
☐	gcmdiso	Local Filesystem	✖	✓	0:1:30	2010-06-14 16:26:03	Edit
☐	landset_etm	Local Filesystem	✖	✓	0:1:30	2010-06-15 02:04:04	Edit
☐	wfs waf	Web DAV	↻	✓	7:0:30	2010-09-17 12:24:42	Edit
☐	wms waf	Web DAV	↻	✓	7:0:0	2010-09-17 11:54:42	Edit
☐	onegeology	CSW/ISO	↻	✓	3:0:0	2010-09-19 12:31:33	Edit

FIGURE 2.4   Timely updating in GEOSS CLH.

2.4   DESIGN AND IMPLEMENTATION OF CLH

2.4.1   ARCHITECTURE

CLH is based on the open-source GeoNetwork software, a standards-based and decentralized spatial information management system designed to enable access to geo-referenced databases and cartographic products from a variety of data providers through descriptive metadata. The development of CLH follows the SOA and Spatial Web Portal. Java Easy Engine for Very Effective Systems (Jeeves) technology (Marsella 2005) is used as the Java engine. It allows the use of XML as an internal data representation and XSL to produce HTML output for clients. The CLH architecture (Figure 2.5) includes three layers: Jeeves layer, GeoNetwork layer, and database layer. Jeeves layer models all requests as XML with XSL transformations. GeoNetwork layer is the core logic layer that provides most system functions

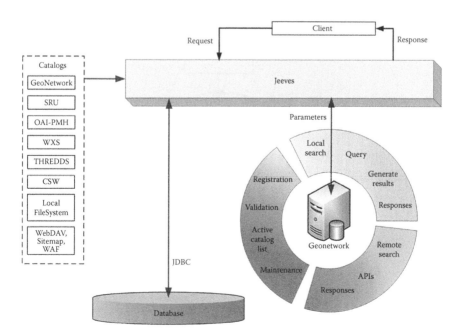

FIGURE 2.5   GEOSS CLH architecture.

like harvesting and search. Database layer is used to store the metadata, and other information (e.g., user information, category information, and group information), MySQL, PostgreSQL, Oracle, or MckoiDB can be used as the database in CLH.

The left of Figure 2.5 are the catalogs from which CLH harvests. The harvesting parameters (e.g., harvesting type, URL of the catalog, periodic harvesting time) are sent to GeoNetwork via Jeeves. A list of service standards is used to get the records in GeoNetwork; for example, the OGC CSW standard (Nebert and Whiteside 2005) is used for CSW catalogs, and the SRU standard (ISO 2010) for SRU catalogs. The metadata are stored in the database, and the index is built after the harvesting. Additionally, users can perform additional operations by sending the request through Jeeves, for example, search metadata, edit metadata, and delete metadata.

2.4.2 ADMINISTRATION, USER, AND GROUP MANAGEMENT

CLH adopts the concepts of user, user group, and user profile to manage registered users and guest users (Figure 2.6). Each group has a number of users with different profiles (e.g., system administrator, user administrator, content reviewer, editor, registered user, and guest), which are stored in a database. Depending on the data owner

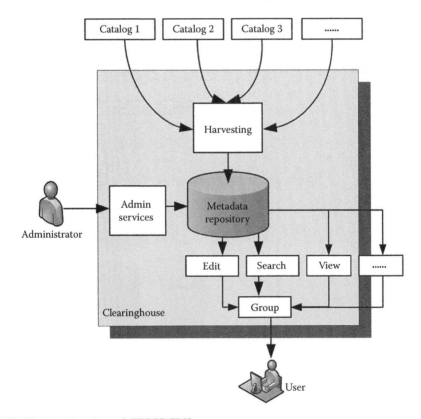

FIGURE 2.6   Functions of GEOSS CLH.

settings, each user group has different data accessibility privileges in relation to view or download metadata and interactive map visualizations for each record in the system.

CLH provides a variety of functions when needed. For example, a metadata registration service is available to allow users to add new metadata to the CLH using the templates if the user has been registered as an editor. CLH provides a set of simplified metadata templates based on the cited standards: ISO, FGDC CSDGM, and DC. CLH also provides editing and deleting operations to update the metadata. These operations are tailored to meet user's privileges. Additionally, CLH provides a default client with an embedded interactive map to visualize WMS.

2.4.3 Harvesting

Harvesting is the process of collecting remote or local metadata and merging them to a local database for faster access. A harvesting entry is usually a periodic schedule of collecting metadata from predefined resources. For example, it can be set to a weekly or daily harvest. Harvesting is not a simple import as local and remote metadata must be kept aligned (Figure 2.7). The parameters will be transported to the harvesting module when one or more harvesting nodes are added and started to

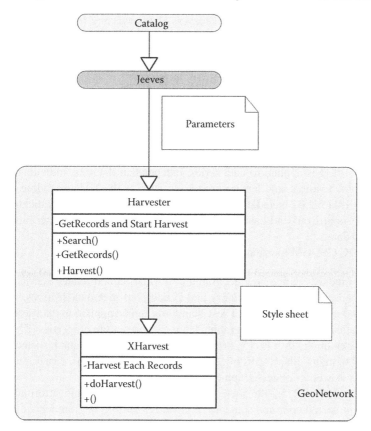

FIGURE 2.7   General workflow of GEOSS CLH harvesting.

harvest through Jeeves. The harvesting functions will be used according to the types of the catalogs. Then each record will be retrieved and transformed to the ISO 19139 if they are nonnative formats supported by the CLH.

CLH supports harvesting through the following protocols, and as of April 1, 2013, the CLH has harvested about 460 K metadata entries:

- Z39.50 *GeoProfile* (manually initiated, under testing)
- CSW 2.0.2 baseline, AP ISO, ebRIM with no extensions
- WebDAV (Whitehead 2010), sitemaps, and Web Accessible Folders (WAFs)
- OGC GetCapabilities (WMS, WFS, WCS) (Vretanos 2002; Evans 2003) endpoints
- Local file access for batch ingest of packaged, static metadata
- THREDDS catalog (Unidata 2010)
- OAI-PMH
- ISO 23950 *SRU*
- GeoNetwork native catalog

The general harvesting workflow of CLH is shown in Figure 2.7. The parameters will be transported to the harvesting module when one or more harvesting nodes are added and started to harvest through Jeeves. The harvesting functions will be used according to the types of catalogs. Then each record will be retrieved and transformed to the ISO 19139, if needed.

2.4.4 METADATA STANDARDS AND TRANSFORMATION

CLH is capable of handling several metadata schemas including the following:

- ISO 19115: GeoNetwork implements a version of the draft, which uses short names for elements.
- ISO 19139 is applied to data series, independent datasets, individual geographic features, and feature properties, such as the XML encoding of the ISO 19115:2007 metadata and ISO 19119 service metadata specifications.
- DC* is utilized and based on a set of elements capable of describing any metadata.
- FGDC CSDGM metadata standards (FGDC n.d.).

The ISO standards define how geographic information and related services should be described and provide mandatory and conditional metadata sections, metadata entities, and metadata elements. These standards can be applied to data series, independent datasets, individual geographic features, and feature properties (Zhang et al. 2010c). For example, ISO 19115:2003 was designed for digital data. Its principles can be applied to many other forms of geographic data such as maps, charts, and textual documents as well as nongeographic data.

With the ISO 19115:2003 metadata standard for geographic information now being the preferred common standard, it is widely used to migrate legacy metadata

---

* http://dublincore.org/

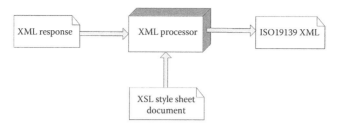

FIGURE 2.8 Metadata transformation.

into the new standard. CLH provides import and export functionalities and has a number of transformers. It is an easy process for a system administrator to install custom transformers based on XSLT (OSGeo 2009). Figure 2.8 illustrates the transformation process: (1) take in an XML response document, (2) apply the XSLT file to it, and (3) generate ISO 19139 XML document.

2.4.5 USER INTERFACE AND PROGRAMMING APIS

2.4.5.1 Search through Web Graphics User Interface

CLH uses Lucene for indexing and searching (Liu et al. 2011). Lucene is a Java library that adds text indexing and searching capabilities to an application. It offers a simple yet powerful core API. By providing a search interface, CLH allows users to search the metadata as needed, including what, where, and when (Figure 2.9). *Title, Abstract, Free Text,* or *Keyword(s)* are used in Lucene index and they can also be used as searching criteria. With bounding box coordinates or country name, a place-based search is supported. Additionally, the created date of metadata, category of metadata, or group can also be used for the search.

2.4.5.2 Remote Search

Besides searching through web Graphics User Interface (GUI), CLH also supports CSW interface* to discover and search EO data. The interface supports the

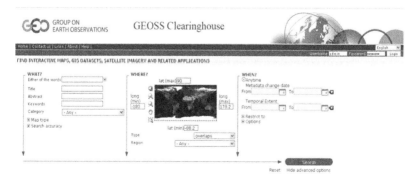

FIGURE 2.9 GEOSS CLH search screen.

---

* http://clearinghouse.cisc.gmu.edu/geonetwork/srv/en/csw

four mandatory CSW requests (GetCapabilities, GetRecords, GetRecordByID, and DescribeRecord). For example, users can use the following GetRecords request to query the total number of CSR component records from the CLH.

```xml
<?xml version="1.0" encoding="UTF-8"?>
<csw:GetRecords
 xmlns="http://www.opengis.net/cat/csw/2.0.2"
 xmlns:csw="http://www.opengis.net/cat/csw/2.0.2"
 xmlns:gmd="http://www.isotc211.org/2005/gmd"
 xmlns:ogc="http://www.opengis.net/ogc"   xmlns:gml="http://www.opengis.
 net/gml"
 xmlns:rim="urn:oasis:names:tc:ebxml-regrep:xsd:rim:3.0"
 service="CSW" version="2.0.2" outputFormat="application/xml"
 outputSchema="http://www.opengis.net/cat/csw/2.0.2" resultType="hits"
 <csw:Query typeNames="csw:Record">
  <csw:ElementSetName>brief</csw:ElementSetName>
  <csw:Constraint version="1.1.0">
   <ogc:Filter>
    <ogc:PropertyIsLike escapeChar="\" singleChar="?" wildCard="*">
     <ogc:PropertyName>AnyText</ogc:PropertyName>
     <ogc:Literal>*csr\:component*</ogc:Literal>
    </ogc:PropertyIsLike>
   </ogc:Filter>
  </csw:Constraint>
 </csw:Query>
</csw:GetRecords>
```

The response of the search is as follows:

```xml
<?xml version="1.0" encoding="UTF-8"?>
<csw:GetRecordsResponsexmlns:csw="http://www.opengis.net/cat/
csw/2.0.2"
xmlns:xsi="http://www.w3.org/2001/XMLSchema-instance"
xsi:schemaLocation="http://www.opengis.net/cat/csw/2.0.2
http://schemas.opengis.net/csw/2.0.2/CSW-discovery.xsd">
 <csw:SearchStatus timestamp="2010-05-06T13:42:06"/>
 <csw:SearchResults numberOfRecordsMatched="236"numberOfRecords
 Returned="0" elementSet="brief" nextRecord="1"/>
</csw:GetRecordsResponse>
```

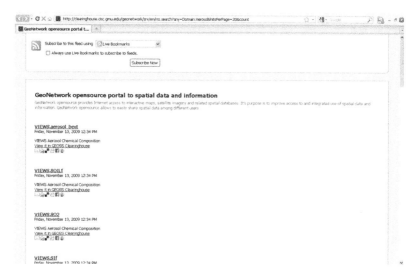

FIGURE 2.10   RSS search results.

In addition, CLH also provides SRU* search and RSS† search API. Figure 2.10 shows the results to get the metadata that contains *Domain:Aerosol.*

2.5   USAGE AND OPERATIONAL STATUS

2.5.1   System Operations

CLH (Figure 2.11) is currently hosted on the Amazon Elastic Cloud Computing (EC2) platform for responding to a large volume of concurrent users from different geographical locations (Huang et al. 2010). We utilized versioning and new technologies (e.g., Web 2.0 AJAX framework) in the development to improve user experience. User communities are encouraged to report the bugs and errors to improve system stability and reliability. Improvements to the CLH will be continually integrated with different GeoNetwork open-source releases.

2.5.2   System Metadata Status

As of April 2013, CLH hosted about 460 K metadata including both metadata for data and catalogs (Figure 2.12). These data have been widely used in different domains such as air quality, agriculture, land cover, and hydrology.

---

* http://clearinghouse.cisc.gmu.edu/geonetwork/srv/en/portal.sru
† http://clearinghouse.cisc.gmu.edu/geonetwork/srv/en/rss.search

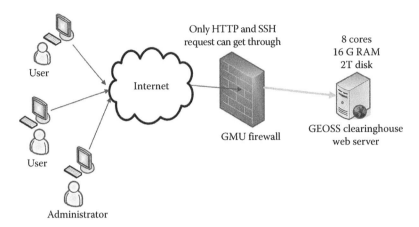

FIGURE 2.11   Operational system of GEOSS CLH.

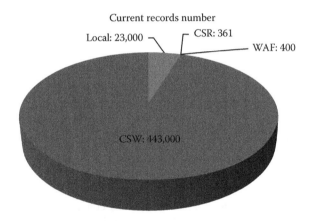

FIGURE 2.12   GEOSS CLH records.

2.5.3   USAGE

One operational function of CLH is to analyze the usage to help improve the operation
in, for example, (1) promoting the usage of the CLH in the area where users are less
explored, (2) using the visiting frequency to identify the popularity in certain regions
for making deployment adjustments, and (3) examining the historical usage changes
through usage data. Usage information is recorded with the access IP and the access
numbers, which can be used to derive the number of users, the frequency of user
access, and the geographical region of the users. We extracted IP information from log
files and associated the IP information with geographical locations using an IP parsing
web service.* End-user distribution and access frequency are depicted in Figure 2.13.

From December 22, 2010 to October 31, 2012, CLH received a total number of
2,202,660 user accesses. The height of the bar is scaled to the visiting frequency

---

* http://api.ipinfodb.com

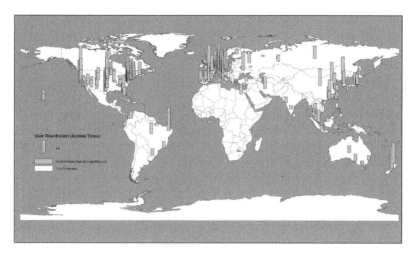

FIGURE 2.13   Usage distribution of GEOSS CLH from December 22, 2010, to October 31, 2012.

(in natural logarithm, Ln). North America, Europe, and East Asia are three high access frequency regions. However, promotion and awareness advertisement should be given in South America, Asia, and Africa to target regions of the world where the highest density of the world's population resides.

2.6   BIG GEOSPATIAL DATA CHALLENGES AND SOLUTIONS

Developing and operating CLH require addressing several big geospatial data challenges including data complexity, distribution of data and catalogs, and performance. These challenges are further discussed in the following.

2.6.1   DATA AND DATABASE

CLH must address challenges of heterogeneity, relationships, and large volume of metadata:

- The number of metadata records is large because new large geospatial data sets are rapidly collected, processed, and published by many organizations on a daily or even hourly basis. How to effectively collect these newly published resources in a timely manner and how to efficiently manage the large amount of collected metadata pose a challenge for CLH. Automatic harvesting (Section 2.4.3) is employed to tackle the collecting challenge; a novel provenance and updating mechanism (Sections 2.3.2 and 2.3.4) is used to address the updating issue. A well-designed database based on PostgreSQL and MySQL is used to store and manage metadata, which are large.
- The metadata are heterogeneous because they are created by various organizations or domain users and published to support specific domain applications, such as climate and polar sciences. These metadata often have different standards, formats, and versions. How to handle the heterogeneity

as well as to keep metadata's integrity for supporting interoperability poses another challenge for CLH. Metadata standardization and transformation (Section 2.4.4) is used to tackle this challenge.

- The relationships among collected metadata are complex. The metadata describe the geospatial resources from different science domains; each record refers to specific geospatial resource. The relationship among these records is complex and not explicitly described and yet is important for delivering the right datasets to the right users especially when the data are shared across different domains. Capturing these relationships and effectively managing them is another challenge for CLH. Semantics and ontology are of potential to address this challenge and being leveraged to optimize the metadata sharing and discovery mechanism (Section 2.2.3). Further studies are required to address how best to collect and utilize semantics into metadata sharing and discovering.

2.6.2 Distribution and Cloud Computing

Two aspects of the big geospatial data distribution need to be considered: (1) CLH harvests the metadata from catalogs that are distributed all over the world and (2) CLH is accessed by end users worldwide.

Since EO data are intrinsically collected and published in a distributed fashion, distributed catalog is one of the big geospatial data challenges when building a global system to integrate, manage, access, and share global big geospatial data. CLH harvests metadata from distributed catalogs periodically; such a harvesting is a time-consuming process that heavily depends on the network connection and the performance of catalog server being harvested. According to spatial principles identified by Yang et al. (2011), deploying the CLH server(s) based on the spatial distribution of the catalog server(s) can improve the harvesting performance.

Intensive concurrent user access can cause slow response by CLH. This issue becomes more challenging with the ever-growing metadata and the increasing number of end users. CLH user access shows a strong spatial pattern where regions with high population densities (e.g., the United States, Europe, and Asia) have a large number of end users and generate lots of accesses (Figure 2.13). Considering the distribution pattern of the end users and utilizing the user access patterns can help improve the system searching performance by deploying multiple CLH servers to the three hot spots (Section 2.5.3). Once the distribution patterns of catalogs and users are identified, (1) spatial principles can be utilized to explore the optimized locations for CLH servers, and (2) cloud computing can be leveraged to deploy multiple server instances to handle the distribution of data and users.

2.6.3 Searching Performance and Index

Data searching is a time-consuming process for CLH in that (1) the big data are spatiotemporal in nature, (2) significant amounts of EO data are produced and registered into CLH on a daily or even hourly basis, and (3) CLH receives large amounts of concurrent data search requests. To improve the searching performance, text indexing and spatial indexing are utilized:

- *Text indexing*: As of April 2013, CLH has harvested 460 K metadata and each metadata store large amounts of text-based information about the data. Searching against large amounts of text could be extremely time consuming without a proper index. Lucene (Section 2.4.5.1) is used in CLH to support text indexing and searching. Title, Abstract, Free Text, or Keyword(s) are indexed in Lucene and they can also be used as searching criteria.
- *Spatial indexing*: CLH supports spatial search of metadata by defining spatial coverage of the search. A spatial index significantly improves CLH searching performance by leveraging spatial relationships among metadata. Spatial features of the metadata in the CLH are indexed using the R-tree supported by PostgreSQL. By leveraging the two indices, CLH provides consistently quick responses (1–2 s) for concurrent searches from various clients (Section 2.3.3).

Table 2.1 summarizes the technical solutions that are used in CLH to address big geospatial data challenges.

TABLE 2.1
Techniques for Addressing Big Geospatial Data Challenges

Techniques/ Challenges	Metadata Collecting/ Managing	Metadata Heterogeneity	Relationship among Metadata	Distribution of Users and Data	Searching Performance
Database	X	—	—	—	X
Automatic harvesting	X	—	—	—	—
Standardization and transformation	X	X	—	—	—
Provenance and updating	X	—	X	—	—
Interoperability	—	X	X	—	—
Buffering/ caching	—	—	—	X	X
Multi-threading	—	—	—	X	q
Cloud computing	X	—	—	X	X
Spatiotemporal indexing	—	—	—	—	X
Spatiotemporal principle	X	—	X	X	X
Semantics/ ontology	X	X	X	—	X

2.7   SUMMARY AND FUTURE RESEARCH

This chapter describes our research and development of CLH to support geospatial resource inventories integration for serving geospatial communities. To develop such an operational system, we addressed four technological challenges: interoperability, metadata provenance, performance, and updating problems. A number of end-user functions are provided to users to configure harvesting, select relevant standards, administer the operational system, conduct search, and perform statistical analyses.

CLH is an evolving operational system with new features analyzed, developed, and integrated continuously. We are currently expanding the interoperability aspect of the system to support more catalogs, testing to connect to portals and other applications, and adding semantic reasoning to support information exchange across domains, cultures, and communities. The global access inspired us to consider deploying the system across different continents through cloud computing and spatiotemporal indexing (Xia 2012). Semantic reasoning-based functionalities are also being added to support meaning-based search. Additionally, we are testing the overall performance to make it a reliable system through well-configured computing platforms for global geospatial communities and evolve it into a kernel component to support the geospatial cyberinfrastructure and geoinformation science (Yang et al. 2010b) over the upcoming decade.

ACKNOWLEDGMENTS

The research and development reported are supported by FGDC and NASA. Ms. Unche A Saydahmat and Mr. Steve McClure proofed this chapter. Ms. Manzhu Yu helped format this chapter.

REFERENCES

Bai, Y. and Di, L. (2011). Providing access to satellite imagery through OGC catalog service interfaces in support of the Global Earth Observation System of Systems. *Computers & Geosciences*, *37*(4), 435–443.

Bishr, Y. (1998). Overcoming the semantic and other barriers to GIS interoperability. *International Journal of Geographical Information Science*, *12*(4), 299–314.

Buneman, P., Khanna, S., and Wang-Chiew, T. (2001). Why and where: A characterization of data provenance. In: Bussche, J. and Vianu, V. (Eds.), *Database Theory—ICDT 2001* (Vol. 1973, pp. 316–330), Berlin, Germany: Springer Berlin Heidelberg.

Christian, E. J. (2005). Planning for the global earth observation system of systems (GEOSS), *Space Policy*, *2*(21), 105–109.

Christian, E. J. (2008). GEOSS architecture principles and the GEOSS clearinghouse. *IEEE Systems Journal*, *2*(3), 333–337.

Craglia, M., Goodchild, M. F., Annoni, A., Camara, G., Gould, M., Kuhn, W. et al. (2008). Next-generation digital earth: A position paper from the Vespucci Initiative for the Advancement of Geographic Information Science. *International Journal of Spatial Data Infrastructures Research*, *3*, 146–167.

Crocker, D. (1982). Standard for the format of ARPA Internet text messages. Retrieved from http://tools.ietf.org/html/rfc822 (Accessed on October 1, 2013).

DCMI (The Dublin Core Metadata Initiative). (2010). Dublin Core. Retrieved from http:// dublincore.org/ (Accessed on September 11, 2010).

Duchossois, G. (2006). The global earth observation system of systems (GEOSS): An example of international cooperation. *2nd ESA/ASI Workshop on International Cooperation for Sustainable Space Exploration Sarteano (Siena)* May 9–12, 2006, pp. 1–29, Siena, Italy.

Evans, J. (2003). Web coverage service (WCS), version1.0.0, OGC 03-065r6. OGC implementation specification. Retrieved from http://portal.opengeospatial.org/files/?artifact_id_3837 (Accessed on May 17, 2013).

FGDC (The Federal Geographic Data Committee). (n.d.). Geospatial metadata standards. Retrieved from http://www.fgdc.gov/metadata/geospatial-metadata-standards (Accessed on March 11, 2013).

GEO (Group on Earth Observations). (2005). The global earth observation system of systems (GEOSS) 10-year implementation plan. The group on earth observations, 11pp. Retrieved from http://www.preventionweb.net/english/professional/publications/v.php?id = 8631 (Accessed on July 02, 2009).

GEO. (2009). The GEOSS common infrastructure. Retrieved from http://www.Earthobservations.org/gci_gci.shtml (Accessed on December 16, 2009).

GEO. (2010). GEOSS portal. Retrieved from http://www.Earthobservations.org/gci_gp.shtml (Accessed on September 11, 2010).

Goldenberg, S. B., Landsea, C. W., Mestas-Nuñez, A. M., and Gray, W. M. (2001). The recent increase in Atlantic hurricane activity: Causes and implications. *Science*, *293*(5529), 474–479.

Huang, M., Maidment, D. R., and Tian, Y. (2011). Using SOA and RIAs for water data discovery and retrieval. *Environmental Modelling and Software*, *26*(11), 1309–1324.

Huang, Q., Yang, C., Nebert, D., Liu, K., and Wu, H. (2010). Cloud computing for geosciences: Deployment of GEOSS clearinghouse on Amazon's EC2. In *Proceedings of the ACM SIGSPATIAL International Workshop on High Performance and Distributed Geographic Information Systems* (pp. 35–38). ACM.

ISO (International Organization for Standardization). (2007). ISO 19139 Geographic information is—Metadata—XML schema implementation. ISO Geneva. Retrieved from http://www.iso.org/iso/catalogue_detail.htm?csnumber=32557 (Accessed on September 11, 2010).

ISO. (2010). Archive of SRU, version 1.1 Specifications. Retrieved from http://www.loc.gov/standards/sru/sru1-1archive/search-retrieve-operation.html (Accessed on September 11, 2010).

Jacobson, M., Charlson, R. J., Rodhe, H., and Orians, G. H. (2000). *Earth System Science: From Biogeochemical Cycles to Global Changes* (Vol. 72), San Diego, CA: Academic Press.

Khaled, R., Tayeb, L. M., and Servigne, S. (2010). Geospatial web services semantic discovery approach using quality. *Journal of Convergence Information Technology*, *5*(2), 28–35.

Khalsa, S. J. S., Nativi, S., and Geller, G. N. (2009). The GEOSS interoperability process pilot project (IP3). *IEEE Transactions on Geoscience and Remote Sensing*, *47*(1), 80–91.

Kuhn, M. (2004). A summary of the international standard date and time notation. Retrieved from http://www.cl.cam.ac.uk/~mgk25/iso-time.html (Accessed on February 13, 2009).

Kusuda, T. and Achenbach, P. R. (1965). *Earth Temperature and Thermal Diffusivity at Selected Stations in the United States*. Gaithersburg, MD: DTIC Document.

Lautenbacher, C. (2006). The global earth observation system of systems: Science serving society, *Space Policy*, *22*(1), 8–11.

Li, W., Yang, C., and Raskin, R. (2008). A semantic enhanced search for spatial web portals. In *AAAI Spring Symposium: Semantic Scientific Knowledge Integration* (pp. 47–50).

Li, W., Yang, C., and Yang, C. (2010). An active crawler for discovering geospatial Web services and their distribution pattern—A case study of OGC web map service. *International Journal of Geographical Information Science*, *24*(8), 1127–1147.

Li, Z., Yang, C., Sun, M., Li, J., Xu, C., Huang, Q., and Liu, K. (2013). A high performance web-based system for analyzing and visualizing spatiotemporal data for climate studies. In: *Web and Wireless Geographical Information Systems* (pp. 190–198). Berlin, Germany: Springer Berlin Heidelberg.

Li, Z., Yang, C. P., Wu, H., Li, W., and Miao, L. (2010). An optimized framework for seamlessly integrating OGC Web Services to support geospatial sciences. *International Journal of Geographical Information Science*, 25(4), 595–613.

Liu, K., Yang, C., Li, W., Li, Z., Wu, H., Rezgui, A. et al. (2011). The GEOSS clearinghouse high performance search engine. Paper presented at the *19th International Conference on Geoinformatics*, June 24–26, 2011, Shanghai, China.

Marsella, M. (2005). *Jeeves Developer's Manual*. Release 1.0, 31pp. Retrieved from https://www.seegrid.csiro.au/wiki/pub/Infosrvices/GeoNetwork/jeeves.doc (Accessed on October 1, 2013).

Nah, F. F.-H. (2004). A study on tolerable waiting time: How long are web users willing to wait? *Behaviour and Information Technology*, 23(3), 153–163.

Nativi, S. and Bigagli, L. (2009). Discovery, mediation, and access services for earth observation data. *IEEE Journal of Selected Topics in Applied Earth Observations and Remote Sensing*, 2(4), 233–240.

Nebert, D. (2004). Developing spatial data infrastructures: The SDI cookbook, GSDI, 2004, pp. 171. Retrieved from http://www.gsdi.org/docs2004/Cookbook/cookbookV2.0.pdf (Accessed on September 06, 2010).

Nebert, D. and Whiteside, A. (2005). Catalog services for the web, version 2, OGC implementation specification. Retrieved from http://portal.opengis.org/files/?artifact_id_5929, OGC (Accessed on October 05, 2010).

NISO (National Information Standards Organization). (2010). Information retrieval (Z39.50): Application service definition and protocol specification. Retrieved from http:// www. loc. gov/z3950/agency/Z39-50-2003.pdf (Accessed on September 11, 2010).

Oppen, D. C. and Dalal, Y. K. (1983). The clearinghouse: A decentralized agent for locating named objects in a distributed environment. *ACM Transactions on Information Systems*, 1(3), 230–253.

OSGeo (Open Source Geospatial Foundation). (2009). GeoNetwork opensource the complete manual. V2.4. Retrieved from http://www.fao.org/geonetwork/docs/Manual.pdf (Accessed on September 11, 2010).

Paul, M. and Ghosh, S. (2008). A service-oriented approach for integrating heterogeneous spatial data sources realization of a virtual geo-data repository. *International Journal of Cooperative Information Systems*, 17(01), 111–153.

Robock, A. (2000). Volcanic eruptions and climate. *Reviews of Geophysics*, 38(2), 191–219.

Simmhan, Y. L., Plale, B., and Gannon, D. (2005). A survey of data provenance in e-science. *ACM Sigmod Record*, 34(3), 31–36.

Ticheler, J. (2008). Open source# 10: Geonetwork opensource. *GEO Connexion*, 7(5), 44–47.

Ticheler, J. and Hielkema, J. U. (2007). GeoNetwork opensource, Internationally standardized distributed spatial information management. *OSGeo Journal*, 2(1). Retrieved from http://journal.osgeo.org/index.php/journal/article/viewFile/86/69

Unidata (2010). THREDDS. Retrieved from http://www.unidata.ucar.edu/projects/THREDDS/ (Accessed on August 10, 2010).

Vretanos, P. A. (2002). Web feature service, version 1.0. Retrieved from http://portal.opengeo-spatial.org/files/?artifact_id_7176, OGC (Accessed on October 05, 2010).

Whitehead, W. (2010). Web DAV. Retrieved from http://www.webdav.org/ (Accessed on September 11, 2010).

Xia, J. (2012). *Optimizing a Spatiotemporal Index to Support GEOSS Clearinghouse*. Master Thesis, Fairfax, VA: George Mason University.

Yang, C., Goodchild, M., Huang, Q., Nebert, D., Raskin, R., Xu, Y., Bambacus, M., and Fay, D. (2011). Spatial cloud computing: How can the geospatial sciences use and help shape cloud computing? *International Journal of Digital Earth*, 4(4), 305–329.

Yang, C., Li, W., Xie, J., and Zhou, B. (2008). Distributed geospatial information processing: Sharing distributed geospatial resources to support Digital Earth. *International Journal of Digital Earth*, 1(3), 259–278.

Yang, C., Raskin, R., Goodchild, M., and Gahegan, M. (2010a). Geospatial cyberinfrastructure: Past, present and future. *Computers, Environment and Urban Systems*, 34(4), 264–277.

Yang, C., Wong, D., Miao, Q., and Yang, R. (2010b). *Advanced Geoinformation Science*, Boca Raton, FL: CRC Press, 485pp.

Yang, C., Wong, D. W., Yang, R., Kafatos, M., and Li, Q. (2005). Performance-improving techniques in web-based GIS. *International Journal of Geographical Information Science*, 19(3), 319–342.

Yue, P., Di, L., Yang, W., Yu, G., and Zhao, P. (2007). Semantics-based automatic composition of geospatial web service chains. *Computers & Geosciences*, 33(5), 649–665.

Zhang, C. and Li, W. (2005). The roles of web feature and web map services in real-time geospatial data sharing for time-critical applications. *Cartography and Geographic Information Science*, 32(4), 269–283.

Zhang, C., Zhao, T., and Li, W. (2010a). Automatic search of geospatial features for disaster and emergency management. *International Journal of Applied Earth Observation and Geoinformation*, 12(6), 409–418.

Zhang, C., Zhao, T., and Li, W. (2010b). The framework of a geospatial semantic web-based spatial decision support system for Digital Earth. *International Journal of Digital Earth*, 3(2), 111–134.

Zhang, C., Zhao, T., Li, W., and Osleeb, J. P. (2010c). Towards logic-based geospatial feature discovery and integration using web feature service and geospatial semantic web. *International Journal of Geographical Information Science*, 24(6), 903–923.

3 Using a Cloud Computing Environment to Process Large 3D Spatial Datasets

Ramanathan Sugumaran, Jeffrey Burnett, and Marc P. Armstrong

CONTENTS

3.1 INTRODUCTION

3.1.1 BIG SPATIAL DATA

Big data is a term in widespread use and refers to datasets that have grown so large that they become difficult to store, manage, share, analyze, and visualize using typical database software tools (White, 2012). According to IBM (2012), big data span four dimensions: volume (data quantity), velocity (real-time processing), variety (multiple sources), and veracity (data accuracy). The tremendous growth of big data during the last few decades is largely a consequence of the increased capacity and plummeting costs associated with computing and information technology.

Geographic information plays a prominent role in the proliferation of big data. It has been estimated that more than 80% of data used by managers and planners are related with spatial or location information (Worral, 1991). During the past three decades, communities around the world have regularly accumulated large collections (multi-petabytes) of 2D and 3D spatial data. Massive 2D spatial data archive examples include repositories of aerial and satellite imagery. For example, the US Geological Survey (USGS) recently announced a 37-year record of global Landsat satellite images (petabytes of data) that have continuously recorded conditions on the Earth's surface. This image resource is now available for free to the public (USGS, 2011). NASA generates roughly 5 TB of remote sensing-related data per day (Vatsavai et al., 2012). In addition to these large 2D image data repositories, a large quantity of 3D spatial data, particularly topographic information, is now increasingly available with the advent of new data capture technologies such as LiDAR.

LiDAR technologies are accurate, cost effective, and timely for elevation data collection when compared with other methods such as photogrammetric techniques (Hodgson et al., 2003; Oryspayev et al., 2012; Sugumaran et al., 2011). This massive quantity of elevation data is valuable to a large number of potential users (academia, local, state, and federal government agencies, tribal governments, environmental and engineering consulting companies), at local to global scales, and can be used to generate high-resolution digital elevation model (DEMs), which in turn can be used to accurately map surface features such as buildings and trees (Meng et al., 2010). As a result, there is a growing number of 3D spatial data applications ranging from urban planning, environmental impact assessment, floodplain mapping, hydrology, geomorphology, forest inventory, urban planning, landscape ecology, cultural heritage protection, transportation management, and disaster preparedness (Anderson et al., 2005; Campbell et al., 2003; Meng et al., 2010; Reutebuch et al., 2005; Tse et al., 2005). For a detailed explanation on advantages and how LiDAR data are collected, refer to Ackermann (1999), Ambercore (2008), Baltsavias (1999), Hodgson et al., (2005), Lohr (1998), Meng et al., (2010), Reutebuch et al., (2005), and Wehr and Lohr (1999).

3.1.2 NEED FOR CLOUD COMPUTING ENVIRONMENT

Though the pace of 3D spatial data collection continues to accelerate, the provision of affordable technology for dealing with issues such as processing, archiving, managing, disseminating, and analyzing large data volumes has lagged (Chen, 2009; Evangelinos and Hill, 2008; Han et al., 2009; Liu and Zhang, 2008; Meng et al., 2010; Oryspayev et al., 2012; Schön et al., 2009; Sugumaran et al., 2011). Of these challenges, dataset size and the computational resources required to process massive amounts of information are major issues for users such as local governments and small businesses. In addition to size, the variety and update rate of datasets often exceed the capacity of commonly used spatial computing and spatial database technologies (Shekhar et al., 2012; Wang et al., 2009; Yang et al., 2011). Processing of large datasets such as LiDAR can be performed using high-performance computing (HPC) environments such as clusters, grid computing, and supercomputers (Simmhan and Ramakrishnan, 2010). However, these platforms require significant investments in equipment and maintenance (Ostermann et al., 2010). A study by Shekhar et al. (2012) addressed the emerging challenges posed by big spatial datasets such as trajectories of cell phones and GPS devices, vehicle engine measurements, and temporally detailed road maps for next-generation transportation routing services. These new challenges include availability of temporally detailed information, growing diversity of big spatial data, need for diverse solution methods, privacy issues surrounding geographic information, and geospatial reasoning across space and time. As a result of these challenges, researchers have begun to search for alternate solutions for the development of large-scale data-intensive applications.

The emergence of cloud computing has brought to the fore a potential solution to this challenge. The cloud computing paradigm is now considered a promising approach for several reasons, including improved performance, *pay-as-you-go* payment regimes, reduced software costs, on-demand provisioning of resources, scalability, elasticity, effectively unlimited storage capacity, device independence, and increased data reliability (Armbrust et al., 2009; Cui et al., 2010; Huang et al., 2010; McEvoy and Schulze, 2008; Rezgui et al., 2013; Watson et al., 2008). Ji et al. (2012) outlined key issues associated with big data processing including big data management platforms and service models, distributed file systems, data storage, data virtualization platforms, and distributed applications in the cloud environment. Though cloud computing services are now available to researchers, only a few investigators have evaluated cloud computing as a viable option for processing large 3D spatial datasets. Krishnan et al. (2010) investigated the implementation of a local gridding algorithm written in C++ along with a cloud-based MapReduce programming model to generate a DEM. In another study, Sugumaran et al. (2011) developed a LiDAR data processing application that uses Amazon EC2 cloud computing with improved data mining algorithms coupled with parallel computing technology. In the next section, we describe the development of cloud computing-based LiDAR processing system (CLiPS), investigate its performance, and analyze the various costs associated with running it using a commercial cloud environment for a statewide Iowa LiDAR database.

3.2   METHODOLOGY

3.2.1   Iowa LiDAR Database

Though the acquisition and use of LiDAR data has grown rapidly during the past decade, only a few countries such as Denmark, Finland, Sweden, and Switzerland either have nationwide LiDAR data or will in the near future. In the United States, the USGS serves as the lead agency in coordinating efforts across multiple agencies towards the development of a national LiDAR dataset. Several states have either collected statewide data or are in the process of doing so. Iowa is one of the states that have completed (in 2010) statewide LiDAR data collection (Iowa DNR, 2009). The raw data for this project consist of more than three terabytes. Currently, the GeoTREE Center serves the statewide LiDAR data for the public through the Iowa LiDAR Portal (ILP) (http://geotree2.geog.uni.edu/lidar/) (Figure 3.1). The data are distributed to the public as a collection of tiles, each of which covers 4 km^2. There are over 34,000 tiles for the entire state.

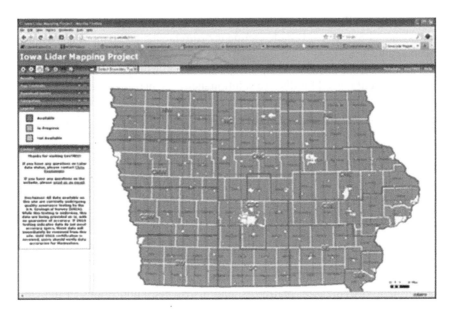

FIGURE 3.1   ILP interface.

The LiDAR data have been widely used by local, state, and private agencies in a number of applications. Figure 3.2 shows the monthly ILP data download from August 2011 to August 2012, with over 25 terabytes of raw data downloaded during this period. The current version of the ILP has several options for data visualization and downloading raw data. Users can interactively select their area of interest or retrieve data based on administrative boundary layers such as cities, counties, or watersheds. However, most users of these data are local and state government agencies and they do not have sufficiently powerful computers to process data for large areas. In addition, the current version of the ILP only allows downloading raw data and provides no additional processing functionality such as creating a DEM and derived products (e.g., slope, aspect, and hill shading) for end users. In order to support these functionalities, the CLiPS system has been developed and implemented using cost-effective cloud computing resources.

In order to implement, process, validate, and verify results, several tiles of the LiDAR topographic data from the Iowa statewide LiDAR project were used. Since tile size varies based on location, the tiles were chosen to represent three different terrain and LiDAR return characteristics: one area with little variation in topography, one with undulating terrain, and one urban area. For each study area, a varying number of tiles ($3 \times 3 = 9$, $5 \times 5 = 25$) were selected for processing. The size of each tile varies from 90 to 377 MB for flat terrain, 137 to 491 MB for undulating terrain, and 181 to 355 MB for urban terrain. Usually, LiDAR points designated as *last returns* are used to create the digital terrain model (DTM) representation of terrain (i.e., bare earth), whereas *first returns* represent the digital surface model (DSM), which includes all features such as trees and buildings. In other cases, *all returns*

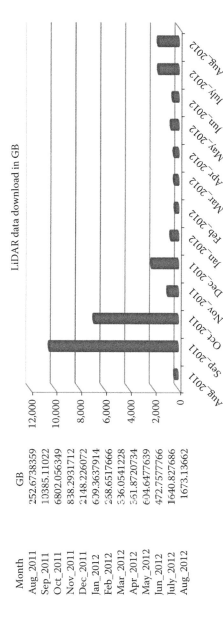

Month	GB
Aug_2011	252.6738359
Sep_2011	10385.11022
Oct_2011	6802.056349
Nov_2011	838.2931712
Dec_2011	2148.226072
Jan_2012	609.3637914
Feb_2012	258.6517666
Mar_2012	536.0541228
Apr_2012	561.8720734
May_2012	604.6477639
Jun_2012	472.7577766
July_2012	1640.827686
Aug_2012	1673.13662

FIGURE 3.2   Monthly data downloads from the statewide LiDAR portal.

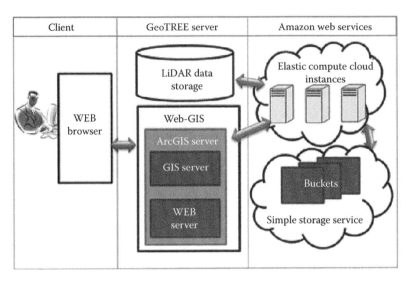

FIGURE 3.3   Overall architecture for CLiPS developed in this study.

refers to a 3D LiDAR cloud in which multiple returns indicate multiple elevations at a location, for example, openings in tree canopies. In our case, we used the large *all returns* datasets to test and validate the approach.

3.2.2   CLiPS Design and Implementation

The overall CLiPS architecture developed in this study is shown in Figure 3.3. The web portal is implemented using Adobe's Flex framework along with Esri's ArcGIS API for Flex (ESRI, 2012). The CLiPS framework was designed and implemented using a three-tier client–server technology (Figure 3.3) that incorporates a user interface in a top tier that supports user interaction with the system, a middle tier that provides process management services such as process development, monitoring, and analysis, and a third tier that provides database management functionality and is dedicated to data and file services. The client-side interface was developed using Flex and the server side uses our own custom built tools, constructed from open-source products such as LasTools (LasTools, 2006), and other spatial server technologies.

3.3   RESULTS

The CLiPS user interface developed in the project is shown in Figure 3.4. Two major groups of functions were developed: preprocessing and postprocessing. At the preprocessing stage, users can invoke a data reduction algorithm (e.g., vertex decimation approach), clip and merge raw data, and initiate user-specified data download tools. The vertex decimation function enables a user to reduce data volume (Oryspayev et al., 2012). During postprocessing, users can choose to create a DEM from a triangulated irregular network (TIN) or point cloud, merge and clip a DEM, and derive secondary products from a DEM or TIN (e.g., generate contours, as well as calculate slope and aspect).

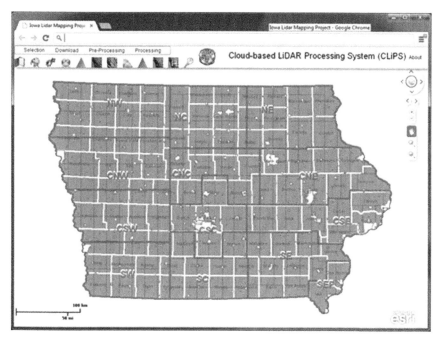

FIGURE 3.4   CLiPS user interface.

TABLE 3.1
EC2 Instant Types Specifications

	Large	XLarge	2XLarge
Memory (GiB)	7.5	15	30
Storage (GB)	850	1690	Elastic block store
I/O performance	Moderate	Moderate	High
Platform	32 bits	32 bits	64 bits

The processing interface also allows the user to either choose a predefined computer infrastructure for the project such as an EC2 instance type (large, Xlarge, and double Xlarge) or specify a configuration of processors specific to their particular problem. Table 3.1 provides high-level specifications for these three instance types. The detailed specifications for each instance such as processor, memory, and storage capacity are given at http://aws.amazon.com/ec2/instance-types/. In order to demonstrate processing flows, an application example is provided in the next section.

3.3.1 APPLICATION EXAMPLE: DEM GENERATION USING LARGE LiDAR DATASETS

In this example, we demonstrate the entire flow of the DEM generation process, starting from how a user selects LiDAR data from the CLiPS web portal to how it is processed at the commercial cloud computing site used (Figure 3.5). In a typical session,

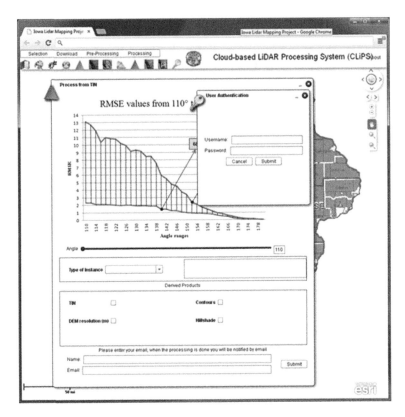

FIGURE 3.5   User interface for generating DEM and derived products.

a user will (1) select a region of interest from the map, (2) provide Amazon Web Services (AWS) credentials to log on to the AWS service, (3) select AWS instance types (computer resources) that will be used, (4) select a preferred processor type (e.g., CPU or GPU), (5) specify the desired products (e.g., TIN, slope, or contour), and (6) specify a name and e-mail address that is used for notification when processing is completed (Figure 3.5). After a user provides the needed information in the user interface on the client side, the selected LiDAR data are pushed from the GeoTREE server to Amazon cloud data storage, then, using the user-specified parameters, a DEM and other derived products are produced using the software we have implemented in the EC2 processing environment. Finally, after processing is completed, an e-mail notification is sent to the user with an ftp location for product download.

In order to test system performance and processing time, we used 18 combinations of datasets and processor configurations: three types of terrain data (flat, undulating, urban), tile sizes (9 and 25), and three EC2 instant types (large, Xlarge, and double Xlarge). Figure 3.6 depicts processing times for these different combinations. As expected, the figure shows that as the computer power increases from large to Xlarge to 2Xlarge, processing time decreases for each

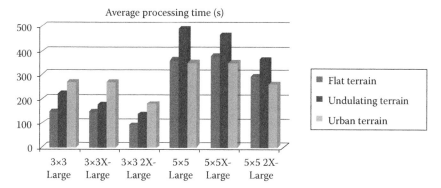

FIGURE 3.6   Comparison of time and performance for different computer resources used in this study.

type of LiDAR data. For the 5 × 5 tile size runs, the undulating terrain data required more processing time than the other two terrain types used. This is due most likely to the variance in height of the points, area covered (50 km²), and the interpolation calculations that are required to generate the digital elevation model. This is not the case in 3 × 3 tile comparison with three terrain types. In this case, urban tiles consistently took more time than the undulating and flat terrains. This may be due to point cloud density and the area covered in the test. These results clearly demonstrate the effect that terrain type and point cloud density has on LiDAR data processing.

The combination that required the largest total processing time was for the large block (5 × 5) with undulating terrain when processed using the smallest processor configuration; it required 8 min and 12 s to merge the tiles and convert them to a DEM. This, however, represents a trade-off since it is the most cost-effective combination, at $34 for a processor hour. The same operation took 4 min and 56 s on the high-memory double extra-large instance. When using the EC2 service, any partial hours are billed as a full hour. Comparisons of various combinations in this study clearly show that even with up to 25 tiles with varying instance types, each request required less than an hour with an associated cost of less than a dollar for data processing. The cost for data upload from the GeoTREE server and data storage on the cloud was less than $50. We did not store the entire statewide data resource using the Amazon cloud storage facility due to the cost involved, though the changing economies of storage may suggest that we revisit this option in the near future. The overall cost for testing the data on the Amazon cloud was less than $100, an amount that is affordable by any of our users.

3.3.2   HEURISTIC MODELS DEVELOPMENT

Based on lessons learned from the different types of instances, tile sizes, terrain types, processing time, and costs involved, we developed two heuristic models to provide novice users with a prototype recommender system: (1) LiDAR data

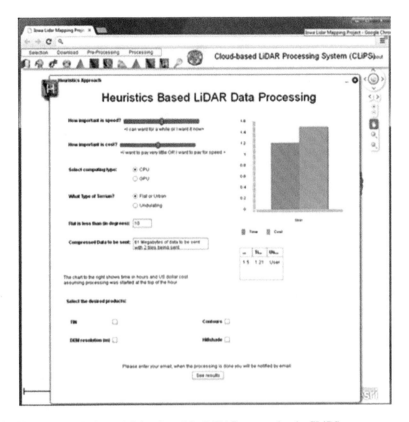

FIGURE 3.7　Heuristic model developed for LiDAR processing in CLiPS.

reduction and (2) best computer infrastructure allocation for various terrain types (Figures 3.5 and 3.7).

The LiDAR data reduction recommender system was developed using a vertex decimation approach to reduce the data density from a TIN (Oryspayev et al., 2012). The vertex decimation approach was originally presented by Schroeder et al. (1992), where every vertex of the TIN model is iteratively checked for its significance to the rendering of the surface model. Based on its significance, the vertex is retained or removed from the TIN model. Each vertex considered for decimation is evaluated with respect to the number of triangles it partly defines and the angular threshold used to determine whether to decimate the vertex. For example, a vertex that serves as the end of an edge for six triangles with a z-value (elevation) that is substantially different than the vertex at the other end of that edge will likely be retained. On the other hand, a vertex that partly defines four triangles and has only a small z-value difference is a candidate for decimation.

In order to develop the heuristic model, the decimation rate for flat and undulating terrain was evaluated using angular ranges in increments of 2 from 110 to 180. Then, for each angle reduction, a DEM was developed and compared with the original

DEM (no reduction) for accuracy estimation. The root mean square error (RMSE) between these two DEMs was computed using the following equation:

$$\text{RMSE} = \sqrt{\frac{\sum_{i=1}^{n} (X_{Original,i} - X_{Obtained,i})^2}{n}}$$

where

> n is the number of cells in a raster
> $X_{Original,i}$ is the original value
> $X_{Obtained,i}$ is the obtained value at cell i

Figure 3.5 shows how the size of the file has been reduced with varying angles used for two terrain types. It also shows how the RMSE value varied based on the size reduction. This test helps users find an optimal angle selection and acceptable RMSE value for their application.

The second recommender system provides users with optimal instance types and processor configurations based on terrain type, how much they are willing to spend, and how quickly they want the job to be completed in the cloud environment. In order to develop the heuristic model, we used LiDAR data from flat and undulating terrain, GPU and CPU configurations on the cloud platform, and expected time and cost for processing the data on the cloud (Figure 3.7). Again the goal is to help novice users specify an appropriate computer configuration that will enable them to process LiDAR data in the cloud and trade-off time and expense.

3.4   CONCLUSIONS

We have successfully developed and implemented a web-based LiDAR data processing system using a cloud computing environment. We tested three types of LiDAR terrain data with different configurations of computer resources using Amazon's EC2 service. The undulating terrain dataset took more time than the other terrain types for 5 × 5 tile groups, while the urban terrain was the most computationally intensive for the 3 × 3 tile groups used in this study. Our results clearly show that as computer power increases, processing times decrease for all three types of LiDAR data. Our approach is scalable in the range of configurations we examined. However, terrain type and point density in the study area are also important factors that influence the processing of LiDAR data. Irrespective of instance type and other factors, the overall cost was modest (less than $100). Our approach clearly demonstrates the advantages that cloud computing environments hold over traditional approaches in time, cost, and performance metrics. In addition, based on lessons learned from the different types of instances, tile sizes, terrain types, processing time, and costs involved, we developed two heuristic models for LiDAR data reduction. These models are incorporated in the user interface and enable novice users to explore suitable computer infrastructure allocations for various terrain types in the cloud environment.

ACKNOWLEDGMENT

This research was conducted with support from Amazon Research Grant and USGS-AmericaView projects.

REFERENCES

Ackermann, F. 1999. Airborne laser scanning-present status and future expectations. *ISPRS Journal of Photogrammetry and Remote Sensing, 542*, 64–67.

Ambercore. 2008. Terrapoint aerial services—A white paper on LIDAR mapping. http://www.ambercore.com/files/TerrapointWhitePaper.pdf, (accessed May 4, 2011).

Anderson, E. S., Thompson, J. A., and Austin, R. E. 2005. LIDAR density and linear interpolator effects on elevation estimates. *International Journal of Remote Sensing, 2618*, 3889–3900.

Armbrust, M., Fox, A., Griffith, R., Joseph, D. A., Katz, R. K., Konwinski, A., Lee, G., Patterson, D. A., Rabkin, R., Stoica, I., and Zaharia, M. 2009. *Above the Clouds: A Berkeley View of Cloud Computing,* Technical Report No. UCB/EECS-2009-28, University of California at Berkeley, Berkeley, CA, February.

Baltsavias, E. P. 1999. A comparison between photogrammetry and laser scanning. *ISPRS Journal of Photogrammetry and Remote Sensing, 542*, 83–94.

Campbell, J., de Haag, M. U., van Graas, F., and Young, S. 2003. Light detection and ranging-based terrain navigation—A concept exploration. In *Proceedings of the 16th International Technical Meeting of the Satellite Division of the Institute of Navigation,* Portland, OR, September 9–12.

Chen, Q. 2009. Improvement of the edge-based morphological EM method for lidar data filtering. *International Journal of Remote Sensing, 304*, 1069–1074.

Cui, D., Wu, Y., and Zhang, Q. 2010. Massive spatial data processing model based on cloud computing model. *Computational Science and Optimization CSO, 2010 3rd International Joint Conference on Computational Science and Optimization, 2*, 347–350.

ESRI. 2012. ArcGIS server. http://www.esri.com/software/arcgis/arcgisserver, (accessed March 23, 2012).

Evangelinos, C. and Hill, C. N. 2008. Cloud computing for parallel scientific HPC applications: Feasibility of running coupled atmosphere-ocean climate models on Amazon's EC2, In *Proceedings of the 1st Workshop on Cloud Computing and Its Applications* CCA, Chicago, IL, October 22–23.

Han, S. H., Heo, J., Sohn, H. G., and Yu, K. 2009. Parallel processing method for airborne laser scanning data using a pc cluster and a virtual grid. *Sensors, 94*, 2555–2573.

Hodgson, M. E., Jensen, J., Raber, G., Tullis, J., Davis, B. A., Thompson, G., and Schuckman, K. 2005. An evaluation of lidar-derived elevation and terrain slope in leaf-off conditions. *Photogrammetric Engineering and Remote Sensing, 717*, 817.

Hodgson, M. E., Jensen, J. R., Schmidt, L., Schill, S., and Davis, B. 2003. An evaluation of LIDAR-and IFSAR-derived digital elevation models in leaf-on conditions with USGS Level 1 and Level 2 DEMs. *Remote Sensing of Environment, 842*, 295–308.

Huang, Q., Yang, C., Nebert, D., Liu, K., and Wu, H. 2010. Cloud computing for geosciences: Deployment of GEOSS clearinghouse on Amazon's EC2, *HPDGIS '10: Proceedings of the ACM SIGSPATIAL International Workshop on High Performance and Distributed Geographic Information Systems,* New York, pp. 35–38.

IBM. 2012. Bringing big data to the enterprise. http://www-01.ibm.com/software/data/bigdata/, (accessed on April 4, 2013).

Iowa DNR. 2009. Retrieved April 29, 2009, from State of Iowa: http://www.iowadnr.gov/mapping/lidar/index.html, (accessed on April 4, 2013).

Ji, C., Li, Y., Qiu, W., Jin, Y., Xu, Y., Awada, U., and Qu, W. 2012. Big data processing: Big challenges and opportunities. *Journal of Interconnection Networks*, *13*(3–4), 1–19. DOI: 10.1142/S0219265912500090.

Krishnan, S., Bary, C., and Crosby, C. 2010. Evaluation of MapReduce for Gridding LIDAR Data, *IEEE 2nd International Conference on Cloud Computing Technology and Science 2010*, pp. 33–40. CloudCom, Indianapolis, IN.

LasTools. 2006. Award-winning software for rapid LiDAR processing. http://www.cs.unc.edu/~isenburg/lastools/, (accessed May 4, 2011).

Liu, X. and Zhang, Z. 2008. LIDAR data reduction for efficient and high quality DEM generation. *International Archives of the Photogrammetry, Remote Sensing and Spatial Information Sciences*, *37*, 173–178.

Lohr, U. 1998. Digital elevation models by laser scanning. *The Photogrammetric Record*, *16*(91), 105–109.

McEvoy, G. V. and Schulze, B. 2008: Using clouds to address grid limitations, In *Proceedings of the 6th International Workshop on Middleware for Grid Computing* MGC, Leuven, Belgium, December 1–5.

Meng, X., Currit, N., and Zhao, K. 2010. Ground filtering algorithms for airborne LiDAR data: A review of critical issues. *Remote Sensing*, *23*, 833–860.

Oryspayev, D., Sugumaran, R., DeGroote, J., and Gray, P. 2012. LiDAR data reduction using vertex decimation and processing with GPGPU and multicore CPU technology. *Computers & Geosciences*, *43*, 118–125.

Ostermann, S., Iosup, A., Yigitbasi, N., Prodan, R., Fahringer, T., and Epema, D. 2010. A performance analysis of EC2 cloud computing services for scientific computing. In *Cloud Computing*, pp. 115–131. Springer Berlin Heidelberg, Munich, Germany.

Reutebuch, S. E., Andersen, H. E., and McGaughey, R. J. 2005. Light detection and ranging LIDAR: An emerging tool for multiple resource inventory. *Journal of Forestry*, *1036*, 286–292.

Rezgui, A., Malik, Z., and Yang, C. 2013. High-resolution spatial interpolation on cloud platforms. In *Proceedings of the 28th Annual ACM Symposium on Applied Computing*, pp. 377–382. ACM, Coimbra, Portugal.

Schön, B., Bertolotto, M., Laefer, D. F., and Morrish, S. 2009. Storage, manipulation, and visualization of LiDAR data. In Fabio Remondino, Sabry El-Hakim, Lorenzo Gonzo, eds. *Proceedings of the 3rd ISPRS International Workshop 3D-ARCH 2009: 3D Virtual Reconstruction and Visualization of Complex Architectures*, Trento, Italy, February 25–28 2009: *International Archives of Photogrammetry, Remote Sensing and Spatial Information Sciences: Volume XXXVIII-5/W1*. International Society of Photogrammetry and Remote Sensing.

Schroeder, W. J., Zarge, J. A., and Lorensen, W. E. 1992. Decimation of triangle meshes. In *ACM SIGGRAPH Computer Graphics*, Vol. 26, No. 2, pp. 65–70. ACM, New York.

Shekhar, S., Gunturi, V., Evans, M. R., and Yang, K. 2012. Spatial big-data challenges intersecting mobility and cloud computing. In *Proceedings of the Eleventh ACM International Workshop on Data Engineering for Wireless and Mobile Access*, pp. 1–6. ACM, Scottsdale, AZ.

Simmhan, Y. and Ramakrishnan, L. 2010. Comparison of resource platform selection approaches for scientific workflows. In *Proceedings of the 19th ACM International Symposium on High Performance Distributed Computing*, pp. 445–450. ACM, Chicago, IL.

Sugumaran, R., Oryspayev, D., and Gray, P. 2011. GPU-based cloud performance for LiDAR data processing. *COM.Geo 2011: 2nd International Conference and Exhibition on Computing for Geospatial Research and Applications*, Washington, DC, May 23–25, 2011.

Tse, R. O. C., Dakowicz, M., Gold, C., and Kidner, D. 2005. Automatic building extrusion from a TIN model using LiDAR and ordnance survey landline data. In *Proceedings GIS Research UK*, *13th Annual Conference*, Glasgow, Scotland, pp. 258–264.

USGS. 2011. Landsat archive. http://landsat.usgs.gov/, (accessed April 7, 2013).

Vatsavai, R. R., Ganguly, A., Chandola, V., Stefanidis, A., Klasky, S., and Shekhar, S. 2012. Spatiotemporal data mining in the era of big spatial data: Algorithms and applications. In *Proceedings of the 1st ACM SIGSPATIAL International Workshop on Analytics for Big Geospatial Data*, pp. 1–10. ACM, New York.

Wang, Y., Wang, S., and Zhou, D. 2009. Retrieving and indexing spatial data in the cloud computing environment. In *Proceedings of the 1st International Conference on Cloud Computing*, Beijing, China, December 1–4, *Lecture Notes in Computer Sciences*, Vol. 5931, pp. 322–331.

Watson, P., Lord, P., Gibson, F., Periorellis, P., and Pitsilis, G. 2008. Cloud computing for e-science with CARMEN. In *2nd Iberian Grid Infrastructure Conference Proceedings*, pp. 3–14. Newcastle-upon-Tyne, U.K.

Wehr, A. and Lohr, U. 1999. Airborne laser scanning—An introduction and overview. *ISPRS Journal of Photogrammetry and Remote Sensing, 54*, 68–82.

White, T. 2012. *Hadoop: The Definitive Guide*, O'Reilly Media, Inc., Sebastopol, CA.

Worrall, L. 1991 *Spatial Analysis and Spatial Policy*, Belhaven Press, London, U.K.

Yang, C., Goodchild, M., Huang, Q., Nebert, D., Raskin, R., Bambacus, M., Xu, Y., and Fay, D. 2011. Spatial cloud computing—How can geospatial sciences use and help to shape cloud computing. *International Journal of Digital Earth, 44*, 305–329.

4 Building Open Environments to Meet Big Data Challenges in Earth Sciences

Meixia Deng and Liping Di

CONTENTS

4.1 INTRODUCTION

Big Data, a popular term emerged with the trend to larger datasets, refers to a collection of datasets so large or complex that it becomes difficult to process using regular database management tools or traditional data processing applications (IDC, 2011; White, 2012; MIKE 2.0). Big Data is posing big challenges and opportunities for every field of the human societies as the *Big Data* era has arrived with the exponential growth, availability, and use of data and information (The Economist, 2010; Reichman, 2011). Big Data does not have to be necessarily big in data size. Big Data issues can be raised either as data volume gets so large and varied (big size of data), or datasets come in all types of formats and from many different sources (high-degree complexity of data), or data are produced and must be processed fast (e.g., near real-time data producing and processing) to meet the demand of information (high velocity of data). The recent discussions on Big Data and how to utilize it as the basis for innovation, differentiation, and growth reflect a wide recognition of Big Data opportunities and challenges.

In fact, the Big Data characteristics are intrinsic in modern Earth sciences (alternatively used with the term *geosciences* in this chapter) and have been identified quite earlier. During the past decades, advances in sensors and data acquisition technology have allowed an increasingly detailed observation of the Earth, fueled the phenomenal growth of remote sensing data, and produced ever-increasing amounts of multisource Earth observation data. The collection of massive Earth observation data leads to the Big Data era in Earth sciences and creates unprecedented opportunities and challenges for Earth sciences (Di and McDonald, 1999; Gore, 1999; Di and Deng, 2005; Craglia et al., 2008; Goodchild, 2008; Deng and Di, 2009). For example, Earth sciences have been evolving into the Earth system science (ESS) concepts and methodologies (Crutzen, 2002). ESS, apart from understanding individual processes of the ocean, land, and atmosphere, recognizes that changes in the Earth are the result of complex interactions among the Earth system components and human activities. It emphasizes the study of the Earth as an integrated system using interdisciplinary approaches instead of traditional approaches (Rankey and Ruzek, 2006). Genetically, ESS is a highly interdisciplinary and data- and compute-intensive science that requires accessing, analyzing, and modeling with Big Earth Data. It plays an essential role in creative and effective use of Big Data for driving efficiency and quality to understand the Earth. In response to the methodological changes, Earth sciences higher education needs to adopt the data-intensive education approach (Deng and Di, 2010a,b). The data-intensive education approach is to train students with data-intensive and compute-intensive methodology and enhanced real-world problem-solving abilities (Deng and Di, 2010b). It embodies many recognized perspectives of education innovation in Earth sciences (or geosciences) education community, including "using extensive remote sensing data in classroom teaching" (Manduca and Mogk, 2002; Ledley et al., 2008), "cultivating technical competence and intellectual self-confidence in research" (Marino et al., 2004), "addressing computational thinking and quantitative skills" (Manduca et al., 2004, 2008), and "engaging student interest and creating understanding through exciting, real-world applications" (Marino et al., 2004). Thus, the data-intensive education approach has a focus on dealing with Big Data collected by Earth observations. It is critical and important for Earth sciences higher education to adopt the data-intensive education approach in order to meet the changing societal demands since today's society needs a large scientific workforce of individuals not only well trained in science disciplines but also mastering data processing and management, computational approaches, software tools, and essential information technology (IT) skills (Deng and Di, 2010b). Only with the data-intensive approach, Earth sciences higher education can keep pace with advances in technologies and take better advantage of the opportunities offered by Big Data and thus produce the workforce needed by the society. In general, modern Earth sciences research, education, and applications all extensively deal with Big Data and highly rely on computer-based IT. Big Data provides profound foundation for new methodologies and paradigm shifts in modern Earth sciences, such as the transformation into ESS from traditional Earth sciences and innovating education with the data-intensive approach. It also helps drive efficiency, quality, and personalized products and services for better understanding the Earth, such as finding hidden relationships and new insights in scientific discovery and exploring work.

Coupled with great opportunities are grand challenges. Typical Big Data challenges are reflected by the multiple layers of difficulties existed in accessing and using Earth sciences data among geosciences research and education communities, such as difficulties in finding, accessing, and using multisource data, lack of analytic functions and computing resources, and very limited data-intensive computing/ modeling capability (Ledley et al., 2008; Lynnes, 2008; Deng and Di, 2009). Big Earth Data problems have severely hampered the Earth sciences or geosciences research, education, and applications. Data users need to put great efforts and time (days, weeks, or even months) in obtaining and preprocessing data into ready-to-use form for analysis. Very often, a scientific task becomes impossible due to the lack of data, computing resources or skills, or due to the requirement of real-time data accessing or modeling capability. Very few educators would like to or are able to adopt the *real* data-intensive approach in their classroom teachings due to the difficulties of accessing and using Earth science Big Data. Instead, most educators might adopt a *pseudo* approach: they prepare some sample data, which are normally selected to only cover a small geographic region, and all students share the same sample datasets for the class exercises. The sample datasets are usually used semester by semester, and students are never exposed to the richness of Earth sciences data and never have the opportunities to learn how to use the vast amount of Earth sciences data in the real-world applications. Even for a *pseudo* approach, educators still need to spend significant amount of time and computing resources to prepare the data and software tools. If a class is big and there are limitations to analysis software licenses, students might need to sign up for computer availability. Consequently, students barely have a chance to learn the data handling and computational skills and cultivate the real-world problem-solving abilities offered by the real data-intensive education approach.

In order to address Big Data issues in Earth sciences, great efforts have been earnestly sought. One major effort is to build open data and information environments that can help meet Big Data challenges. Numerous international organizations or societies, such as the Open Geospatial Consortium (OGC), the International Society of Digital Earth (ISDE), and the Committee on Data for Science and Technology (CODATA), have largely encouraged and promoted this effort. Many US federal agencies, such as NSF, NASA, and NOAA, have invested significant funds in supporting the effort. The Center for Spatial Information Science and Systems (CSISS) at George Mason University, supported financially by a number of US federal agencies and international organizations, has dedicated to building a series of open data and information environments during the past decade to address various Big Data issues in Earth sciences research, education, and applications. For example, CSISS has built the following four open data and information environments:

1. GeoBrain (http://geobrain.laits.gmu.edu), an open and data-intensive cyberinfrastructure dedicated to facilitating Earth sciences higher education and research. GeoBrain is the first operational geospatial web service system developed in the world, addressing the barriers that hamper effective access to, use of, and modeling with large volumes of Earth sciences data through an innovative geospatial web service approach (Di, 2004; Di and Deng, 2005; Deng and Di, 2006b, 2008, 2009, 2010a,b). It particularly mobilizes NASA Earth Observation System (EOS) data, making enormous EOS data

easily accessible, usable, and analyzable anytime and anywhere with any Internet-connected computer. It largely facilitates Earth sciences higher education and research with a data-intensive online learning and research environment.

2. CropScape (http://nassgeodata.gmu.edu/CropScape), an interactive web cropland data layer (CDL) exploring system to effectively and efficiently meet the growing needs of instant access to and analysis of crop data and information. Based on the GeoBrain cyberinfrastructure, CropScape is able to provide comprehensive capabilities in online and on-demand querying, visualization, dissemination, and analysis of the US Department of Agriculture (USDA) CDL data. It not only offers the online functionalities of interactive map operations, data customization and downloading, crop acreage statistics, charting and graphing, and multi-temporal change analysis in an interoperable manner but also provides web geoprocessing services such as automatic data delivery and on-demand crop statistics for uses in other applications (Han et al., 2012a). It largely facilitates delivery and analysis of geospatial cropland information for decision support and various research endeavors.

3. VegScape (http://dss.csiss.gmu.edu/VegScape/), a national crop progress monitoring system leveraging GeoBrain functionalities, sensor web technology, and advanced remote sensing technologies to effectively support agricultural decision-makings with critical geospatial crop development progress information. VegScape aims at overcoming the shortcomings of existing crop progress systems/procedures (Yu et al., 2009, 2012a,b). Based on GeoBrain, advanced workflow technology for composing web services was used in automatically producing and disseminating an objective and quantitative national crop progress map and associated decision support data at high-spatial resolution, as well as summary reports.

4. Global Agricultural Drought Monitoring and Forecasting System (GADMFS) (http://gis.csiss.gmu.edu/GADMFS/), a next-generation agricultural drought data and information system that provides near real-time monitoring, on-demand analysis, and forecasting of global agricultural drought. GADMFS, built on the GeoBrain cyberinfrastructure, specifically addresses the Big Data issues related to access and visualization, customization, dissemination, and analysis of historical and near real-time global agricultural drought data and information (Deng et al., 2012, 2013). GADMFS meets the increasing demand on near real-time global agricultural drought monitoring, analysis, and prediction services and complete agricultural drought data and information from diverse users including decision-makers, educators, researchers, farmers, students, and the general public.

The four instances mentioned earlier are able to provide a glimpse of how to meet Big Data challenges in Earth sciences through building open data and information environments. This chapter aims to present an overview of technology foundations and the general system development methodology and approaches for building the four open environments and share experiences, lessons learned, and visions for addressing Big Data challenges in Earth sciences through demonstrations of these four instances developed by CSISS.

4.2 TECHNOLOGY FOUNDATION AND METHODOLOGY

The open environments built by CSISS, despite their different foci on specialized Earth science research, education, and application areas, involve accessing, processing, and analyzing large volumes of satellites data (Big Earth Data). Thus, they need to meet imminent Big Data challenges (e.g., data visualization and analysis) through addressing fundamental issues in Big Data accessibility, interoperability, and usability. They also need to meet the challenges of geoprocessing modeling with Big Data since most geospatial scientific products are not obtained directly from measurements but derived from other data through geoprocessing modeling that consists of constructing and executing a geospatial process model. Geoprocessing modeling is an important approach to produce user-required geospatial scientific products (e.g., derived data and information products). Figure 4.1 illustrates the three phases from a geospatial process model to a user-required scientific product through geoprocessing modeling: knowledge capture phase, user query phase, and user retrieval phase. In the knowledge capture phase, a geospatial process model is constructed, conceptually leading to the geoprocessing steps to generate a type (or types) of scientific product. A geospatial process model always represents behavior-based high-level geospatial knowledge, which can be created in one

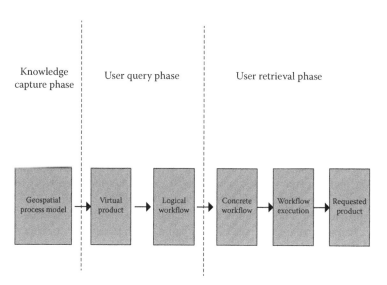

FIGURE 4.1 Three phases of geoprocessing modeling. (From Di, L., *Proc. IEEE Int. Geosci. Remote Sens. Symp.*, 6, 4215, 2005.)

of two ways: manually by a domain expert or automatically by a system (the system needs to have domain knowledge and artificial intelligence capabilities). In the user query phase, a geospatial process model is checked to see if it can be logically instantiated, and if does, a virtual product and a logical workflow are generated. A virtual product is a data or information product not existing in the system but the system knows how to generate it on demand. In the user retrieval phase, a logical workflow is physically instantiated as a concrete workflow and executed to generate a user-required product upon user's request. Many challenges remain in the phases of geoprocessing modeling.

To address the challenges of Big Data and geoprocessing modeling, the open environments need to address three key technical issues at the very least: (1) how to enable better interoperability at data, functions (e.g., geoprocessing services), systems, and semantics levels; (2) how to support Big Data discovery, integration, visualization, analysis, and modeling; and (3) how to obtain an integrated infrastructure to meet the common needs of research, education, and applications by comprehensively addressing Big Data issues in Earth sciences.

This section will present an overview of the technology foundation and general system methodology for addressing the key issues and removing major barriers to use of Big Earth science Data in building the four open environments.

4.2.1 INTEROPERABILITY

Interoperability is the first key issue that needs to be addressed in building open environments. Goodchild (1992) and Bishr (1998) discussed the importance and trend of interoperability for geographic information science and systems and major barriers to interoperability. Barriers to geospatial interoperability can be classified into different types (e.g., organizational, cultural, and technical) and different levels (e.g., data, service, system, and semantics). A fundamental reason causing Big Earth science Data issues is that current geospatial data and information systems are not easily, adequately, or properly interoperable at the data, functions (services), systems, and semantics levels due to technological, organizational, and cultural barriers. As a result, a key technical challenge for building an effective Earth sciences open data and information environment to address Big Data issues is how to enable better interoperability at data, functions (services), system, and semantics levels. In order to address this key challenge, CSISS has put significant efforts in developing new IT such as geospatial web service technology based on existing and ongoing technology and research results such as OGC geospatial interoperability technical specifications (Deng et al., 2003, 2004; Deng and Di, 2009, 2010a).

OGC provides profound technical foundation for enabling geospatial interoperability. It has developed a set of geospatial interoperability specifications in compliance with related international standards, for example, International Standard Organization (ISO) standards. OGC defines interoperability specifications for protocols and interfaces that enable disparate geospatial datasets and maps to be exchanged (served) over the Internet in such a way that conforming clients can integrate multiple products into a single geographically consistent product. In the four open environments developed by CSISS, OGC technical specifications for geospatial interoperability are well observed and the following OGC key specifications adopted:

1. Web coverage service (WCS) specification, which defines standard interfaces for coverage servers so that client applications can seamlessly query and access raw or processed coverage data stored on one or more distributed servers (WCS, 2012)
2. Web map service (WMS) specification, which provides standard interfaces for web clients to uniformly access maps rendered by map servers on the Internet (WMS, 2012)
3. Web feature service (WFS) specification, which describes feature manipulation operations for servers and clients to communicate at the vector feature level (WFS, 2012)
4. Catalogue service for web (CSW) specification, which defines common interfaces that enable diverse applications to perform discovery, browse, and query operations against spatial data metadata and services metadata through the web (CS, 2012)
5. Web processing service (WPS) specification, which provides interface rules that standardize inputs and outputs (i.e., requests and responses) for geospatial processing services (WPS, 2012)

In addition to the earlier OGC specifications, a set of World Wide Web Consortium (W3C) and Organization for the Advancement of Structured Information Standards (OASIS) standards, such as web service description language (WSDL), simple object access protocol (SOAP), and web services business process execution language (WS-BPEL), are also observed and utilized (SOAP, 2012; WSBPEL, 2012; WSDL, 2012).

In brief, the desired data-, service-, semantics-, and system-level interoperability are enabled in the open environments through two major efforts:

1. Design and implementation of a common data environment (CDE) for providing standard interfaces for discovering and accessing diverse data and information in distributed data archives and repositories. A set of OGC data discovery and access specifications, including WCS (Whiteside and Evans, 2008), WFS (Vretanos, 2005), WMS (de la Beaujardière, 2006), and CSW (Martel, 2009), have been implemented together with the CSW-based catalogue federation service (CFS) (Bai et al., 2007) to form CDE for federated discovery of and access to distributed data and information.
2. Design and implementation of a common service environment (CSE) for both human and machine users to discover and invoke geoprocessing services (analytic functions) at a standard way for processing and analyzing data obtained from CDE. A set of standard interface protocols, including WSDL, SOAP, WPS, WS-BPEL for web service and service chaining, and augmented CSW and CFS with ontology (Bai et al., 2009) for service discovery and registry, have been implemented for CSE.

The multilevel interoperability enabled through the CDE and CSE approaches mentioned earlier also helps address data accessibility and usability issues.

With CDE and CSE, data and services in distributed systems can be accessed and used over Internet from a single point of entry through machine-to-machine interfaces.

4.2.2 SERVICEABILITY

The complexities of Earth sciences data cause big gaps between data users' needs and data providers' capabilities. Bridging these gaps requires providing better Big Data discovery, integration, visualization, analysis, and modeling services so that data users could easily access, understand, and use the data provided. Problems still remain because data in most current systems are not interoperable and lack sufficient services to handle them. In order to support better Big Data discovery, integration, visualization, analysis, and modeling, an open environment should further address data interoperability, accessibility, and usability through improved serviceability (i.e., personalized serviceability). The personalized serviceability of the open environments, built in CSISS, is enabled, in addition to the earlier addressed multilevel interoperability, through coupling data with sufficient data, information, and knowledge services.

Sufficient data services mean that any data available in the open environments are coupled with adequate metadata and data transformation services (e.g., subsetting, reformatting, resampling, and re-projection) so that the data can be easily discovered, accessed, visualized, customized, retrieved, and integrated. For instance, providing customized data at user's request through developing interoperable, personalized, and on-demand data access and services (IPODAS) is an initial approach to sufficient data services (Deng and Di, 2009). IPODAS, first developed in GeoBrain and consists of a series of OGC data services (e.g., WCS, WFS, and WMS) and data transformation services (e.g., web coordinate transformation service), effectively eliminates some preliminary difficulties in accessing, integrating, and using geospatial data.

Sufficient information services mean that online, on-demand, and near real-time geospatial analysis of data can be performed to generate user-required information products. Geospatial data, such as remote sensing data, are mostly Big Data with multispectral or even hyper-spectral images. They have large dimensions, with lots of spectral bands (e.g., over two hundreds of bands), far from the *readability* of other data types, such as text and numbers. Geospatial analysis of remote sensing Big Data relies on proper computer-based techniques. Interoperable geospatial web services provide great possibilities and flexibility for geospatial processing and geospatial analysis tasks in distributed environments (Deng et al., 2003, 2004; Deng and Di, 2009, 2010a). In order to enable sufficient information services in the open environments, all data analysis functions are provided as standards-compliant, interoperable, and dynamically chainable geospatial web services. A large number of value-added geospatial web services have been implemented for on-demand geospatial analysis catering for users' specifications. In particular, nearly 80 geospatial web services converted from wrapping functions in the Geographic Resources Analysis Support System (GRASS) (Deng and Di, 2008, 2010a; Li et al., 2010) are made available to worldwide users for manipulating and processing raster, vector, and satellite image data. These web services are standards-compliant web services supporting HTTP GET and POST requests. Each service is described with WSDL and the service input and output are accessible network points, that is, uniform resource locators (URLs), and can be discovered over the Internet. As a result, these geospatial web services could be easily invoked and used by any user for on-demand geospatial processing and analysis. Enabling sufficient information services through this approach will dramatically facilitate scientific studies that need to use geospatial Big Data and will ultimately free users from worrying about the lack of computing resources or skills.

Sufficient knowledge services require automated geoprocessing modeling capability, which enables a system to understand and respond to human and machine queries and generates high-level data, information, and knowledge products (e.g., answers to questions). Sufficient knowledge services are not fully implemented in the open environments yet due to grand challenges, but research progress has been largely made, such as the ones in Di (2005), Deng and Di (2006a), Yue et al. (2006, 2007), and Di et al. (2007). With current technology, rudimentary knowledge services can be provided through manipulated geoprocessing modeling enabled by mechanisms of construction, instantiation, and execution of a geospatial process model (Chen et al., 2009b; Deng and Di, 2010a). In the open environments developed by CSISS, geo-tree is used to describe geoprocessing modeling conceptually (Di and McDonald, 1999; Di, 2005; Deng and Di, 2010a), as shown in Figure 4.2.

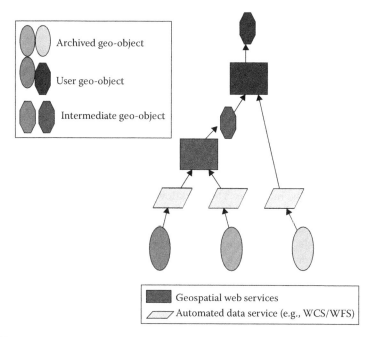

FIGURE 4.2 Geoprocessing modeling with a geo-tree. (From Di, L., *Proc. IEEE Int. Geosci. Remote Sens. Symp.*, 6, 4215, 2005; Deng, M. and Di, L., *Facilitating Data-Intensive Research and Education in Earth Science—A Geospatial Web Service Approach*, Saarbrücken, Germany, 2010a.)

A geo-tree graphically represents a geospatial process model and a type of virtual product that can be generated on demand. There are two types of nodes in a geo-tree: process node and geo-object node. The process node is a geospatial service type. The geo-object node contains a geospatial product (e.g., data) type. A user-requested geospatial product type is named as a user object. A user object can be decomposed into a geo-tree, as in Figure 4.2, consisting of a series of process nodes, intermediate objects, and archived geo-objects or other existing user objects. Thus, the root node in a geo-tree is a user object and a virtual geo-object type that the geospatial process model can generate. An intermediate geo-object in a geo-tree is also a virtual product that can be decomposed into the subordinate geo-tree. The leaf node of a geo-tree (the lowest-level node) is not a virtual product but the type of a real product (e.g., an archived geo-object or existing user object) physically located in some system. A virtual product can be instantiated with proper data and service instances and executed to generate many actual information or knowledge products. With the scheme of a geo-tree, a geospatial process model can be viewed as an abstract workflow describing geoprocessing steps to generate a specific type of geospatial product, not involving in any real data or service instances, can be encoded in a workflow language such as BPEL, and can be saved (i.e., registered) as a virtual product in a system (Chen et al., 2009b; Deng and Di, 2010a). By defining a geospatial process model at the abstract level through a geo-tree, a general model can be developed and used for generating unlimited numbers of instances. Also, this scheme allows domain experts to create models easier by eliminating the needs to details on product specifications.

A workflow engine is needed to execute a geoprocessing workflow encoded in BPEL for generating the desired result. A workflow engine can have three major functions: invoking and executing a geospatial web service workflow (chained services) and generating the result as requested, working as a virtual product manager that helps requestors find both data instances and data types, and instantiating and executing an instantiated geo-tree or workflows. BPELPower (Deng and Di, 2010a; Yu et al., 2012c) is the workflow engine developed to serve all the open environments in CSISS. The engine is an internal service requestor for individual geospatial web services. It will push the input data to the service, execute the service, and pull out the output from the service. If the geo-information requested is a virtual product, the engine will manage the execution of the workflow and deliver the materialized geo-information to the requestor. The workflow engine will also manage the status of the workflow execution and the temporal storage. The execution of a geoprocessing workflow generated from a geospatial process model converts raw observation data to high-level data, information and knowledge products required by the users. A system will become more knowledgeable and capable when more geospatial process models are included in it.

With enhanced serviceability enabled by sufficient data, information, and knowledge services, an open environment is able to address major Big Earth Data challenges and support discovery, integration, visualization, analysis, and modeling of Big Data. For example, the challenges for integrating geospatial data from multiple sources can be addressed through a workflow-based approach: first, developing both data and service ontologies to semantically describe the data and services and then defining a data-integration workflow as a sequence of services where, for each adjacent pair of services, occurrence of the first action is necessary for the occurrence of the second action. A data-integration workflow can be executed through a workflow engine to automatically generate the integrated data.

4.2.3 INFRASTRUCTURE

In order to meet the common needs of geosciences research, education, and applications, the infrastructure of an open environment must be able to comprehensively address Big Data interoperability, accessibility, and usability issues. In other words, the infrastructure of an open environment needs to be interoperable with other infrastructures or systems and needs to be integrated for making distributed data, services, and computational resources easily available, accessible, and usable online. Doing so will free users from worrying about the lack of data and computing resources, save users significant time and effort in discovering and acquiring data, and simplify some complicated scientific tasks. In order to make distributed data, information, and computing resources available at large organizations easily accessible and usable, the design and implementation of the open environment infrastructure in CSISS observe the following principles:

1. *Interoperability*: The system has the capability of plug-in and play over the web so that related geospatial resources (e.g., data, services, service systems) from other sources can be plugged into the system on the fly for meeting the on-demand needs of a wider, more diverse audiences of both human and machine users (the system is interoperable at data, service, semantics, and system levels).
2. *Scalability*: The system is scalable from the construction perspective (e.g., component-based and modular) and user-orientation perspective (e.g., able to deal with any scale from local to global per users' requests).
3. *Reusability*: The system is reusable at function, component, and system levels.
4. *Adaptability*: The system is flexible, extensible, self-evolvable, and maintainable.

Observing these principles, the open environments built in CSISS share a common infrastructure: a three-tier, standards-compliant, and service-oriented architecture (SOA)-based geospatial web service system, as illustrated in Figure 4.3.

The front-end tier is the presentation and distribution component/tier and includes the web portals and clients, which can access the system functionalities and disseminate data and information to users. The web portals in this tier provide end users very convenient online access to all data and information products available through the system, virtual products registered in the system, and the geospatial analysis and geoprocessing/modeling capability enabled by the system. In addition to the web portals, standards-compliant clients can be developed to access all system functionalities as stand-alone applications in this tier to provide some functional privileges

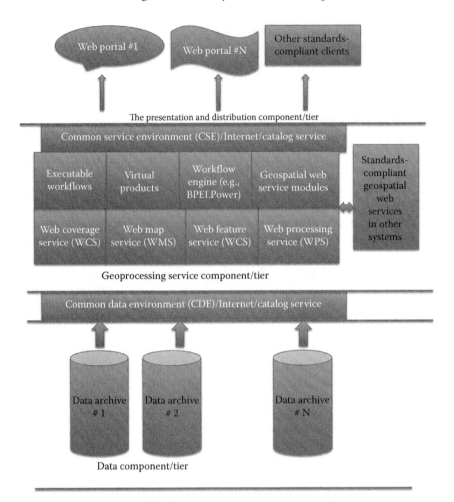

FIGURE 4.3　Architecture of a geospatial web service system.

and offline processing capability to users. This tier, through CSE, provides access to all real and virtual data and information products in the system and all other integrated OGC-compliant providers.

The middle tier is the geoprocessing service component/tier, which consists of multiple components that perform geospatial data processing, information extraction, geoprocessing modeling, on-demand analysis, and knowledge management services. This tier also contains all executable files for individual standards-compliant geospatial web services—self-contained, self-describing, modular applications that can be published, located, and dynamically invoked across the web. Once a service is deployed, other applications (and other web services) can discover and invoke that deployed service. All geospatial web services in the system are developed following the applicable OGC web service standards including OGC WMS, WCS, WFS, WPS, and CSW. They will act as individual *building blocks* for dynamically constructing the complex geospatial models in the system.

At the back end is data component/tier allowing the system to access distributed data resources regardless of where or how the data are archived as long as the data-provider services comply with the interface standards. This tier consists of data servers using the defined CDE to provide data to the data requestors seamlessly regardless of the format, projection, resolution, and the archive location of data. The data requestors of the system include system's geoprocessing service middleware, any application and data analysis systems, application clients, and human users. The CDE allows requestors to specify the requirements for the data they want. A CDE-compliant server must preprocess the data on demand given the requirements and then return the data back to the requestor in the form specified by the requestor.

In summary, the four open environments in CSISS share the common infrastructure of an interoperable and extensible geospatial web service system with the following characteristics: (1) accessible online—users only need to have an Internet-connected PC with a web browser to conduct complicated tasks needing distributed computing resources; (2) standards-based data and services access—the system provides a single point of entry to distributed data and services through standards-compliant common data and service environment; and (3) web service-based data access, integration, analysis, and geoprocessing modeling—all analytic functions are provided as interoperable and chainable geospatial web services, allowing users to easily integrate data and perform online and on-demand geoprocessing modeling and geospatial analysis. Such a common infrastructure enables easy access to, use of, and modeling with distributed Big Earth sciences Data and massive information from a single point of entry such as a web browser.

4.3   DISCUSSIONS

The four instances of open data and information environments in CSISS, GeoBrain, CropScape, VegScape, and GADMFS, have very different foci, functionalities, usages, application areas, and targeted audiences. As mentioned earlier, GeoBrain dedicates to innovating higher education and research in Earth sciences with the data-intensive approach and meeting a wide-range computational needs of Big Earth Data users, while the other three aim to provide specialized agricultural data and information for supporting effective and timely agricultural monitoring and decision-making in different applications. However, they have all been designed and implemented as open, three-tier, component-based, standards-compliant, and SOA-based geospatial web service systems. Why do they share the common infrastructure of such a geospatial web service system? In order to answer this question, the fundamental functionality requirements for open environments need to be mentioned first. Clearly, all the four open data and information environments, despite their specialized applications, must address major Big Data issues for (1) organizing and delivery of large volumes of datasets; (2) disseminating data, information, and model results quickly to a wide audience; and (3) enabling thorough, collaborative, and on-demand analysis to users' best advantages. They also need to highlight the role of data-intensive science in transforming raw observations into applicable, intelligible results and discoveries. Thus, the open environments share the following common functionality features: (1) enabled easy access to, integration, and use of distributed geospatial resources; (2) provided online and on-demand data visualization, analysis, and modeling services; (3) enhanced data-intensive and distributed computing capabilities; and (4) strengthened interoperability among components and systems. To provide these features, an open, three-tier, standards-compliant, component-based, and SOA-based geospatial web service system seems to be the best choice due to its many advantages for distributed computing, such as providing better portability and flexibility (Deng et al., 2004; Deng and Di, 2010a).

There are great benefits in implementing open data and information environments with the addressed common infrastructure of an open, three-tier, standards-compliant, component-based, and SOA-based geospatial web service system. Such a system is generically interoperable, extensible, reusable, scalable, self-evolvable, and maintainable. The three-tier system structure can significantly reduce the requirements for computing resources at the users' side and provide users most conveniences. For example, users of the open environments can perform complex tasks anytime anywhere with any Internet-connected computer because they can use a web browser to access all system functionalities with the infrastructure. Standards-compliance helps address better interoperability, reusability, and multipurpose functionalities provided by individual geospatial web services and their composed workflows. Component-based design and implementation help improved maintenance of the system since a component's failure will not affect the other components or the system functionality and also increases the possibility of reusing the system and components. GeoBrain's infrastructure and functions (e.g., data and geoprocessing services) are reused in all other three systems, and functions developed in CropScape are also reused in VegScape and GADMFS. An SOA-oriented structure allows quick response to new requirements for geospatial data and information and assimilates new geospatial data and processing services from service providers (Granell et al., 2010). Adopting SOA also maximizes scalability for future system augmentation, and as a result, many specialized applications can be easily added to the infrastructure. For instance, GeoBrain has been augmented with many specialized applications, such as the Geospatial Online Analysis System (GeOnAS, 2012; Zhao et al., 2012) and the customized digital elevation model (DEM) data explorer (DEMExploer, 2012; Han et al., 2012b), to meet different user needs. In general, the geospatial web service system approach helps best leverage recent advances in cyberinfrastructure (NSF CIC, 2007), and web services and geospatial interoperability technologies can save significant efforts and computational resources during the implementations of the open environment in CSISS.

In general, all the four environments meet the fundamental functionality requirements of enabling easy access to, use of, and modeling with distributed Big Earth sciences Data and massive information from a single point of entry (e.g., a web browser) through the geospatial web service approach. They are able to remove various difficulties in Big Earth Data, including (1) accessing distributed geospatial data over different data centers with diverse data access protocols; (2) integrating heterogeneous geospatial data due to the format, projection, and coverage complexities; (3) making computing resources available to most users to process, analyze, or model data; and (4) having complete information that supports decision-making, research, and applications. GeoBrain, the first operational geospatial web service system developed in the

world, has well addressed the fundamental requirements and removed difficulties in Big Earth Data. New applications can be easily built on GeoBrain.
In fact, CropScape, VegScape, and GADMFS all are applications built on
GeoBrain, just like the previously mentioned GeOnAS and DEMExploer but
using quite different data sources and information services. GeoBrain provides fundamental infrastructure, abundant resources, and fruitful technology
(Di, 2004; Deng and Di, 2006b, 2008, 2009; Di and Deng, 2010), particularly
geospatial interoperability, web processing, and workflow services (Di et al.,
2005, 2007; Yue et al., 2006; Li et al., 2010; Yu et al., 2012c), as well as the sensor web technology (Chen et al., 2009a; Di et al., 2010; Zhao et al., 2010; Zhang
et al., 2012), for developing an open data and information environment. Reusing
the existing services and tools has saved significant efforts and resources for
building all the open environments in CSISS, which embodies the essence and
benefits of a standards-compliant geospatial web service system like GeoBrain.

Interoperability should be particularly emphasized in building any open
environment. As discussed earlier, a standards-compliant approach has been adopted
for enabling interoperability in building all the open environments in CSISS, which
proves to be very successful. The authors of this chapter consider it as the best option
so far for enabling wide interoperability in building open data and information environments, despite some existing arguments about the costs vs. benefits of adopting standards. The geospatial interoperability technology developed in GeoBrain
has been successfully used for building other open data and information environments in CSISS. However, it still remains a grand challenge to enable adequate and
proper interoperability among different data, services, and systems due to technical,
cultural, and organizational barriers.

Although the four environments are all designed and implemented as comprehensive and extensive cyberinfrastructure aiming to provide sufficient data, information, and knowledge services, they are only implemented to provide sufficient data
and information services operationally at this stage (knowledge services are very
rudimentary, as discussed in Section 4.2.2). Even for data and information services,
they may be *sufficient* at one time but then become *insufficient* at another time due
to new requirements and changing needs. For better responses to new requirements
and changing needs, two issues need to be taken into careful consideration: service
modularity and community involvement. Service modularity is an important factor
affecting flexibility, applicability, and reusability of service modules for different
geospatial process models (Deng and Di, 2010a). If a service module is too small,
many modules are needed to construct a complex geospatial process model, hence
affecting system performance. If a service module is too big, the module will not be
easily reused for different geospatial process models. Thus, the designed modularity of geospatial web services has a direct impact on the system response to the new
requirements and needs and must be carefully considered. The community involvement is another significant factor for building effective open environments because
the real power of a geospatial web service system relies on the availability of a large
number of chainable geoprocessing service modules and geospatial process models.
Building a geoprocessing service module or geospatial process model from scratch
is very time consuming and needs specific domain knowledge. Building an adequate

number of geoprocessing service modules or geospatial models to meet diverse requirements and needs is beyond the ability of any individual or even a development team. The community-wise or across-community efforts and contributions need to be largely encouraged for the system implementation and evolution of an open environment in order to provide sufficient data and information services for meeting the ever-changing requirements and demands.

At present, the implementation of sufficient knowledge services is very much behind the demand. From the open environments perspective, users can dynamically get on-demand and customized data products or derived information products through sufficient data and information services or rudimentary knowledge products through manipulated geoprocessing modeling instead of raw data, but the system could not automatically generate the knowledge products to answer users' queries. Significant research efforts are needed to provide sufficient knowledge services and make the open environments an intelligent geospatial knowledge system that can more effectively support decision-makings. An intelligent knowledge system is a system that can provide user-specific knowledge to respond to a user's query, with some fundamental characteristics, such as automatic reasoning mechanisms, dynamic knowledge generation, and question-answering capabilities. In such a system, intelligence is characterized as the system's capability to understand a user's question and to find the right answer to the question. If such an answer does not preexist, the system should be able to generate an answer from existing lower-level data or information. Therefore, an intelligent knowledge system automatically converts data to user-specific geospatial information and knowledge, which is the answer to the user-specified question. Such a system should be able to answer many *what if* questions by automatically and intelligently chaining individual service instances to form a complex model matching the input data with the model and executing the model to deliver the answer user. Four aspects of the geospatial question-answering mechanism in a web service-based geospatial knowledge system need to be addressed: (1) standards-based automated geospatial data and services discovery and access; (2) intelligent decomposition driven by domain knowledge and ontology of a user query into a geospatial process model for workflow construction; (3) automated geospatial web service chaining, binding, and execution of the workflow; and (4) management of workflows and geospatial models.

Building an intelligent geospatial knowledge system based on current open data and information environments will be a significant leap forward helping Earth sciences to solve numerous challenging scientific problems that threaten the health and safety of human beings and supporting decision-makers for selecting better policies. The study of functional requirements, methodology, and system architecture in building the open environments in CSISS has considered the framework, mechanisms, and service tools needed for geoprocessing modeling and knowledge building and sharing, but full implementation of sufficient knowledge services and an intelligent knowledge system has not been completed yet. Besides, the study of the quality of service (QoS) of individual geospatial web services from different providers and mechanism for measurement of QoS is not fully considered either, although the service metrics have been recorded for all the open environments with performance tests using software tools such as JMeter (Deng et al., 2013; JMeter, 2013). It is logical to put future research efforts on implementing sufficient knowledge services, improving QoS, and developing more effective open environments with sufficient and intelligent data, information, and knowledge services.

4.4   SUMMARY

Modern Earth sciences exploration and discovery activities mainly rely on data-intensive geospatial computing and modeling capability. Rapid climate and environmental changes place new requirements for geospatial computing and modeling capability in understanding, predicting, and responding to these changes effectively. Due to the grand Big Earth Data challenges, current data-intensive geospatial computing and modeling capability are still largely limited. Building open and data-intensive information and knowledge environments to address the confronting Big Data challenges in Earth sciences is of great importance in providing reasonable solutions to the problems we face with climate change, environment protection, and other global issues. CSISS at GMU has put significant efforts in building a series of open data-intensive environments for meeting wide-range data and information needs in Earth sciences research, education, and applications. Four instances of the environments, GeoBrain, CropScape, VegScape, and GADMFS, are presented in this chapter to exemplify the general methodology and approaches for building an open, data-intensive information and knowledge environment in helping Earth sciences scientific and engineering activities and providing societal benefits. All those environments have been built through an innovative geospatial web service approach. The operational uses of those environments among worldwide users have demonstrated that a geospatial web service approach can effectively build open environments meeting various demands and building such environments is a very practical approach to addressing solutions to Big Data issues in Earth sciences. The open environments enable easy access to, analysis of, and modeling with distributed Big Earth Data and computing resources through a single point of entry and provide innovative methods for automatically generating customized data, information, and knowledge products. This will largely facilitate and promote Earth sciences research, education, and application, since it is very expensive or impossible for individual scientists, educators, or students to have all the required computing resources and computer skills to conduct all data-intensive research and education activities due to the complexities involved. In addition, open data, information, and knowledge environments can be expected to play a critical role in promoting scientific democracy progress: the general public can freely access vast data, information, and computing resources (which were only available to a few privileged scientists before) to conduct scientific investigations based on their personal interests.

The open environments built in CSISS are able to provide experiences, lessons learned, technology readiness, and insightful vision to develop more effective data, information, and knowledge environments for addressing the increasing

Big Data challenges and wider-range geospatial computing and modeling needs of future generations of scientists, researchers, educators, students, and the general public.

REFERENCES

Bai, Y., L. Di, A. Chen, Y. Liu, and Y. Wei, 2007. Towards a geospatial catalogue federation service, *Photogrammetric Engineering and Remote Sensing*, 73(6), 699–708.

Bai, Y., L. Di, and Y. Wei, 2009. A taxonomy of geospatial services for global service discovery and interoperability, *Computers & Geosciences*, 35(4), 783–790.

Bishr, Y., 1998. Overcoming the semantic and other barriers to GIS interoperability, *International Journal of Geographic Information Science*, 12(4), 299–314.

Craglia, M., Goodchild. M., Annoni, A., Camara, G., Gould, M., Kuhn, W., Mark, D., and Masser, I., 2008. Next-generation digital earth, *International Journal of Spatial Data Infrastructure Research*, 3, 146–165.

Crutzen, P. J., 2002. The anthropocene: Geology of mankind, *Nature*, 415, 23.

Chen, A., L. Di, Y. Wei, Y. Bai, and Y. Liu, 2009b. Use of grid computing for modeling virtual geospatial products, *International Journal of Geographic Information Science*, 23(5), 581–604.

Chen, N., L. Di, G. Yu, and M. Min, 2009a. A flexible geospatial sensor observation service for diverse sensor data based on web services, *ISPRS Journal of Photogrammetry and Remote Sensing*, 64(2), 234–242, doi:10.1016/j.isprsjprs.2008.12.001.

CS (Catalogue Service), 2012. OpenGIS® catalogue services specification, (version 2.0.2, OGC 07-006r1). Open Geospatial Consortium Inc., D. Nebert, A. Whiteside, and Panagiotis (Peter) Vretanos (Eds.), 2007-02-23.

de la Beaujardière, J. (Ed.), 2006. OpenGIS Web Map Server Implementation Specification (OGC 06-042, Version 1.3.0). Open Geospatial Consortium Inc., Wayland, MA, 85pp. Available at http://portal.opengeospatial.org/files/?artifact_id = 14416, accessed March 23, 2010.

DEMExplorer, 2012. DEM Explorer. Available at http://ws.csiss.gmu.edu/DEMExplorer/, accessed April 21, 2013.

Deng, M. and L. Di, 2006a. A prototype intelligent geospatial knowledge system based on semantic web service technology, *The International Conference of Geoinformatics 2006*, Wuhan, China.

Deng, M. and L. Di, 2006b. Utilization of latest geospatial web service technologies for remote sensing education through GeoBrain system, *Proceedings of IEEE International Geoscience and Remote Sensing Symposium (IGARSS 2006)*, Denver, CO.

Deng, M. and L. Di, 2008. GeoBrain online resources for supporting college-level data intensive geospatial science and engineering education, *Proceedings of IEEE International Geoscience and Remote Sensing Symposium (IGARSS) 2008*, Boston, MA.

Deng, M. and L. Di, 2009. Building an online learning and research environment to enhance use of geospatial data, *International Journal of Spatial Data Infrastructures Research*, 4, 77–95.

Deng, M. and L. Di, 2010a. *Facilitating Data-Intensive Research and Education in Earth Science—A Geospatial Web Service Approach*. LAP LAMBERT Academic Publishing GmbH & Co. KG, Saarbrücken, Germany.

Deng, M. and L. Di, 2010b. GeoBrain for data-intensive earth science (ES) education. In: C. Yang, D. Wong, Q. Miao, and R. Yang (Eds.) *Advanced Geoinformation Science*, p. 351. CRC Press.

Deng, M., L. Di, W. Han, A. Yagci, C. Peng, and G. Heo, 2013. Web-service-based monitoring and analysis of global agricultural drought, *Photogrammetric Engineering and Remote Sensing*, 79(10), 929–943.

Deng, M., L. Di, G. Yu, A. Yagci, C. Peng, B. Zhang, and D. Shen, 2012. Building an on-demand web service system for global agricultural drought monitoring and forecasting, *Proceedings of 2012 IEEE International Geoscience and Remote Sensing*, Munich, Germany, July 22–27. 4pp.

Deng, M., Y. Liu, and L. Di, 2003. An interoperable web-based image classification service for remote sensing data, *Proceedings of 2003 Asia GIS Conference, Asia GIS Society*, October 16–18, Wuhan, China, 11pp. (CD-ROM publication).

Deng, M., P. Zhao, Y. Liu, A. Chen, and L. Di, 2004. The development of a prototype geospatial web service system for remote sensing data, The International Archives of Photogrammetry, Remote Sensing, and Spatial Information Sciences, http://www.isprs.org/congresses/istanbul2004/comm2/papers/126.pdf.

Di, L., 2004. GeoBrain-A web services based geospatial knowledge building system, *Proceedings of NASA Earth Science Technology Conference 2004*, June 22–24, 2004, Palo Alto, CA, 8pp. (CD-ROM).

Di, L., 2005. Customizable virtual geospatial products at web/grid service environment, *Proceedings of IEEE International Geoscience and Remote Sensing Symposium (IGARSS)*, vol. 6, pp. 4215–4218.

Di, L. and M. Deng, 2005. NEHEA and GeoBrain—An organization and system for data-intensive earth system science education and research at colleges around the world, *Presentations at International Society for Global Spatial Data Infrastructure and International Federation of Surveyors*, Greenbelt, MD, April 2005.

Di, L. and M. Deng, 2010. Enhancing remote sensing education with geobrain cyberinfrastructure, *Geoscience and Remote Sensing Symposium (IGARSS), 2010 IEEE International*, DOI: 10.1109/IGARSS.2010.5650462, pp. 98–101.

Di, L. and K. McDonald, 1999. Next generation data and information systems for earth sciences research, *Proceedings of the First International Symposium on Digital Earth*, Volume I Science Press, Beijing, China, pp. 92–101.

Di, L., K. Moe, and L. van Zyl Terence, 2010. Earth observation sensor web: An overview, *IEEE Journal of Selected Topics in Applied Earth Observations and Remote Sensing*, 3(4), 415–417, doi:10.1109/JSTARS.2010.2089575.

Di, L., P. Yue, W. Yang, G. Yu, P. Zhao, and Y. Wei, 2007. Ontology-supported automatic service chaining for geospatial knowledge discovery, *Proceedings of Annual Meeting of American Society of Photogrammetry and Remote Sensing*, Tampa, FA, May 7–11, 2007, American Society of Photogrammetry and Remote Sensing, Bethesda, MD, unpaginated CD-ROM.

Di, L., P. Zhao, W. Yang, G. Yu, and P. Yue, 2005. Intelligent geospatial web services, *Proceedings of IEEE International Geoscience and Remote Sensing Symposium (IGARSS)* 2005, Seoul, South Korea, July 25–29, pp. 1229–1232.

The Economist, 2012. Data, data everywhere. February 25, 2010. http://www.economist.com/node/15557443, accessed April 10, 2012.

GeOnAS, 2012. An SOA based online geospatial analysis system. Available at http://geobrain.laits.gmu.edu/OnAS/, accessed April 21, 2013.

Goodchild, M. F., 1992. Geographic information science, *International Journal of Geographic Information Systems*, 6(1), 31–45.

Goodchild, M. F., 2008. The use cases of digital earth, *International Journal of Digital Earth*, 1(1), 31–42.

Gore, A., 1999. The digital earth: Understanding our planet in the 21st Century, *Photogrammetric Engineering and Remote Sensing*, 65(5), 528.

Granell, C., D. Díaz, and M. Gould, 2010. Service-oriented applications for environmental models: Reusable geospatial services, *Environmental Modelling and Software*, 25(2), 182–198.

Han, W., L. Di, P. Zhao, and Y. Shao, 2012b. DEM explorer: An online interoperable DEM data sharing and analysis system, *Environmental Modelling & Software*, 38, 101–107.

Han, W., Z. Yang, L. Di, and R. Mueller, 2012a. CropScape: A web service based application for exploring and disseminating US conterminous geospatial cropland data products for decision support, *Computers and Electronics in Agriculture*, 84, 111–123.

IDC, 2011. Big data analytics: Future architectures, skills and roadmaps for the CIO, September 2011.

JMeter, 2013. The Apache JMeter desktop application. Available at http://jmeter.apache.org/index.html, accessed February 20, 2013.

Ledley, T. S., A. Prakash, C. Manduca, and S. Fox, 2008. Recommendations for making geoscience data accessible and usable in education, *EOS*, 89(32), 291.

Li, X., L. Di, W. Han, P. Zhao, and U. Dadi, 2010. Sharing geoscience algorithms in a web service-oriented environment (GRASS GIS example), *Computers & Geosciences*, 36(8), 1060–1068, August 2010. doi:10.1016/j.cageo.2010.03.004.

Lynnes, C., 2008. *Earth Science Data Usability Vision, A Document Recorded after a Discussion during 7th NASA Earth Science Data Systems Working Group Meeting in 2008*, Philadelphia, PA.

Manduca, C. and D. Mogk, 2002. *Using Data in Undergraduate Science Classrooms*, Final Report on an interdisciplinary workshop held at Carleton college, Northfield, MN, April 2002. http://serc.carleton.edu/files/usingdata/UsingData.pdf.

Manduca, C., D. Mogk, and N. Stillings, 2004. *Bringing Research on Learning to the Geosciences*. Report from a workshop sponsored by the NSF and the Johnson Foundation, Racine, WI. http://serc.carleton.edu/files/research_on_learning/ROL0304_2004.pdf.

Manduca, C. A., E. Baer, G. Hancock, R. H. Macdonald, S. Patterson, M. Savina, and J. Wenner, 2008. Making undergraduate geoscience quantitative, *EOS*, 89(16), 149–150. http://serc.carleton.edu/serc/EOS-89-16-2008.html.

Marino, M., T. Sumner, and M. Wright, 2004. *Geoscience Education and Cyberinfrastructure*. Report from a workshop held in Boulder, Colorado, Boulder, CO, April 19–20.

Martel, R. (Ed.), 2009. OpenGIS CSW-ebRIM registry service—Part 1: ebRIM profile of CSW (OGC 07-110r4, Version 1.0.1). Open Geospatial Consortium Inc., Wayland, MA. Available at http://portal.opengeospatial.org/files/?artifact_id=31137, accessed February 15, 2013.

Mike 2.0. Big data definition, http://mike2.openmethodology.org/wiki/Big_Data_Definition, accessed April 10, 2013.

NSF CIC, 2007. *Cyberinfrastructure Vision for 21st Century Discovery*. National Science Foundation Cyberinfrastructure Council, Arlington, VA, March 2007. http://www.nsf.gov/pubs/2007/nsf0728/index.jsp.

Rankey, E. C. and M. Ruzek, 2006. Symphony of the spheres: Perspectives on earth system science education, *Journal of Geoscience Education*, 54(3), 197–201.

Reichman, O. J., Jones, M. B., and Schildhauer, M. P., 2011. Challenges and opportunities of open data in ecology, *Science,* 331(6018), 703–705. doi:10.1126/science.1197962.

SOAP (Simple Object Access Protocol), 2012. Latest SOAP versions. Available at http://www.w3.org/TR/soap/, accessed October 22, 2012.

Vretanos, P. A. (Ed.), 2005. OpenGIS web feature service implementation specification (OGC 04-094, Version 1.1.0). Open Geospatial Consortium Inc., Wayland, MA, 131pp., available at http://www.opengeospatial.org/standards/wfs, accessed October 25, 2012.

WCS (Web Coverage Service), 2012. OpenGIS web coverage service (WCS) implementation standard (OGC 07-067r5, Version 1.1.2). Open Geospatial Consortium Inc., A. Whiteside and J. D. Evans (Eds.), 2008. Available at http://www.opengeospatial.org/standards/wcs, accessed October 20, 2012.

WFS (Web Feature Service), 2012. OpenGIS web feature service implementation specification (OGC 04-094, Version 1.1.0), Open Geospatial Consortium Inc., P. A. Vretanos (Ed.), 2005. Available at http://www.opengeospatial.org/standards/wfs, accessed October 20, 2012.

White, T., 2012. *Hadoop: The Definitive Guide*. O'Reilly Media, (May 10, 2012).

Whiteside, A. and J. D. Evans (Eds.), 2008. OpenGIS Web Coverage Service (WCS) implementation standard (OGC 07-067r5, Version 1.1.2). Open Geospatial Consortium Inc., Wayland, MA, 133pp. Available at http://portal.opengeospatial.org/files/?artifact_id = 27297, accessed March 25, 2011.

WMS (Web Map Service), 2012. OpenGIS® web map server implementation specification, (Version: 1.3.0, OGC® 06-042), Open Geospatial Consortium Inc., 2006-03-15, J. de la Beaujardiere (Ed.). Available at http://www.opengeospatial.org/standards/wms, accessed October 14, 2012.

WPS (Web Processing Service), 2012. OpenGIS® web processing service, (Version: 1.0.0, OGC 05-007r7), Open Geospatial Consortium Inc., P. Schut, (Ed.), 2007-06-08, available at http://www.opengeospatial.org/standards/wps, accessed October 14, 2012.

WSBPEL (Web Services Business Process Execution Language), 2012. Web services business process execution language version 2.0. OASIS standard, 04-11-2007. Available at http://docs.oasis-open.org/wsbpel/2.0/OS/wsbpel-v2.0-OS.html, accessed October 14, 2012.

WSDL, 2012. Web Services Description Language (WSDL) 1.1, W3C note 15 March 2001. Available at http://www.w3.org/TR/wsdl, accessed October 10, 2012.

Yu, G., L. Di, Z. Yang, Y. Shen, Z. Chen, and B. Zhang, 2012b. Corn growth stage estimation using time series vegetation index, *Proceedings of 2012 First International Conference on Agro-Geoinformatics (Agro-Geoinformatics 2012)*, pp. 1–6, doi:10.1109/Agro-Geoinformatics.2012.6311631.

Yu, G., L. Di, Z. Yang, Z. Chen, and B. Zhang, 2012a. Crop condition assessment using high temporal resolution satellite images, *Proceedings of 2012 First International Conference on Agro-Geoinformatics (Agro-Geoinformatics 2012)*, pp. 1–6, doi:10.1109/Agro-Geoinformatics.2012.6311629.

Yu, G., Z. Yang, and L. Di, 2009. Web service based architecture for US national crop progress monitoring system, *Proceedings of IEEE Geoscience and Remote Sensing Symposium*, *(IGARSS 2009)* 4, pp. IV-789–IV-792, doi:10.1109/IGARSS.2009.5417495.

Yu, G., P. Zhao, L. Di, A. Chen, M. Deng, and Y. Bai, 2012c. BPELPower—A BPEL execution engine for geospatial web services, *Computer & Geosciences*, 47, 87–101.

Yue, P., L. Di, W. Yang, G. Yu, and P. Zhao, 2006. Path planning for chaining geospatial web services, *Lecture Notes in Computer Science*, 4295, 214–226.

Yue P., L. Di, W. Yang, G. Yu, and P. Zhao, 2007. Semantics-based automatic composition of geospatial web services chains, *Computers & Geosciences*, 33(5), 649–665, May 2007.

Zhang, B., L. Di, G. Yu, W. Han, and H. Wang, 2012. Towards data and sensor planning service for coupling earth science models and earth observations, *IEEE Journal of Selected Topics in Applied Earth Obervations and Remote Sensing*, 10.1109/JSTARS.2012.2195639.

Zhao, P., L. Di, W. Han, and X. Li, 2012. Building a web-services based geospatial online analysis system, *Journal of Selected Topics in Applied Earth Observations and Remote Sensing*, 5(6), 1780–1782.

Zhao, P., L. Di, and G. Yu, 2010. Toward autonomous mining of the sensor web, *Annals of Information Systems*, 11, 289–307, doi: 10.1007/978-1-4419-5908-9_8.

5 Developing Online Visualization and Analysis Services for NASA Satellite-Derived Global Precipitation Products during the Big Geospatial Data Era

Zhong Liu, Dana Ostrenga, William Teng, and Steven Kempler

CONTENTS

5.1 INTRODUCTION

Precipitation is an important weather variable affecting our daily lives. Each year, severe droughts and floods happen around the world and often cause heavy property damages and human fatalities (Obasi 1994). An example is the tropical cyclone Nargis that made a landfall in Burma on May 2, 2008, causing catastrophic destruction and claiming at least 138,000 lives (http://www.unisdr.org/archive/8742). Accurate measurement and prediction of precipitation in both short and long terms can greatly help hazard preparedness and mitigation efforts. However, precipitation is difficult to measure in data sparse oceans, continents, and remote regions, such as Africa, due to the lack of rain gauge and ground radar networks, creating an obstacle for monitoring applications and forecasting verification. Satellite remote sensing takes an important role to measure precipitation from space on a global scale (Special Issue on TRMM 2000) and there have been many ongoing activities especially in the following areas: product development, product uncertainty, research using precipitation products, and applications.

In the past decade, satellite-derived products have provided a cost-effective way to measure precipitation from space and fill in data gaps in regions mentioned earlier. In recent years, algorithms that utilize multisatellites and multisensors that consist of microwave sensors, geostationary infrared (IR) sensors, and gauges provide both research and near real-time global precipitation products, such as the Tropical Rainfall Measuring Mission (TRMM) (Special Issue on TRMM 2000) Multi-Satellite Precipitation Analysis (TMPA) product suite (Huffman et al. 1995, 1997, 2001, 2007, 2010) and others (Aonashi et al. 2009; Behrangi et al. 2009; Hong et al. 2007; Joyce et al. 2004; Mahrooghy et al. 2012; Sorooshian et al. 2000). These products are widely used in hydrometeorological research and applications. Traditional ways of acquiring and analyzing raw precipitation data products can be a tedious and often time consuming, sometimes a frustrating, experience due to various factors (e.g., data format, data structure, data volume, data subsetting, software development). Online-based tools can provide a quick and easy way for accessing remote sensing precipitation products because downloading data and software is not often needed (Liu et al. 2007, 2009, 2012).

In reality, satellite-derived precipitation products are not perfect, and issues, such as bias, uncertainty, sampling, and sensor calibration, exist (Chiu et al. 2006; Chokngamwong and Chiu 2008; Habib et al. 2012; Rozante et al. 2010; Tian ct al. 2010; Tian and Peters-Lidard 2010; Yilmaz et al. 2010). It is important to quantify these issues in order for users to apply the knowledge gained to their research and applications. On the other hand, as algorithms are improved and newer versions of products are released, changes in product characteristics need to be well understood in order for users to make adjustments or changes to their applications and research. The traditional way of acquiring this knowledge relies on journal publications that, however, usually provide a general overview and seldom address issues and changes at local or regional scales. In addition, the approach often takes time for the results to be published, can be tedious and time consuming, and sometimes a frustrating experience if the user wants to conduct their own assessment from scratch, that is, from raw data, due to various factors (e.g., data format, data structure, data volumes, data subsetting, and new software development). Again, online-based tools can overcome these issues and provide simple access to this information because downloading data and software is not often required (Liu et al. 2009).

Precipitation is one of the key variables (temperature, pressure, wind, cloud, precipitation, humidity, air density) in weather and climate. Precipitation is frequently used in weather and climate studies. An example is the Year of Tropical Convection (YoTC), a joint activity of the *World Climate Research Programme (WCRP)* and *World Weather Research Programme (WWRP)/The Observing System Research and Predictability Experiment (THORPEX)*. It is a 2 year period (May 2008–April 2010) of coordinated observing, modeling, and forecasting with a focus on organized tropical convection, its prediction, and predictability (Moncrieff et al. 2012; Ostrenga et al. 2010). The lack of fundamental knowledge and practical capabilities in tropical convection leaves us disadvantaged in modeling and predicting prominent phenomena of the tropical atmosphere such as the intertropical convergence zone (ITCZ), the El Niño southern oscillation (ENSO), monsoons and their active/break periods, the Madden–Julian oscillation (MJO), subtropical stratus decks, near-surface ocean properties, tropical cyclones, and even the diurnal cycle. A major obstacle in research and model improvement of tropical convection and its multiscale structure is that there is currently no single web portal or application that provides *one-stop shopping* for the relevant heterogeneous National Aeronautics and Space Administration (NASA) multisatellite, model assimilation, and ground observation data. Obtaining satellite data for activities such as case study, product evaluation, model verification, and uncertainty investigation for similar parameters, among other activities, can be a daunting experience due to complicated data structure, data format, software installation, etc. Nonetheless, it is a challenge to develop data services to address these difficulties.

Data services to support precipitation applications are important for maximizing the NASA TRMM and the future Global Precipitation Measurement (GPM) missions' societal benefits. A wide range of social and economic issues are related to weather and climate. An example is that Petherick (2012) used rainfall data to examine the relation between rainfall variability and gross domestic product (GDP) growth in a rain-fed economic country, Ethiopia. Petherick (2012) found that the GDP growth in Ethiopia had a close relationship with rainfall variability after analyzing the rainfall and GDP data between 1980 and 2000; high GDP growth during above average rainfall and low growth during below average rainfall that is normally associated with drought. In the 2007 National Research Council's Decadal Survey (Anthes et al. 2007), it is recommended that the development of an Earth knowledge and information system should consider socioeconomic factors. In short, fast and timely available data access and availability are among important factors for ensuring successful applications.

The NASA Goddard Earth Sciences Data and Information Services Center (GES DISC) is home to the TRMM data archive (Liu et al. 2012; Vicente and the GES DAAC Hydrology Data Support Team, 2007). Since the launch of TRMM in November 1997, the GES DISC has developed tools and services that facilitate precipitation data information access for the precipitation community and other communities as well (Liu et al. 2007, 2010, 2011, 2012; Vicente and the GES DAAC Hydrology Data Support Team, 2007). For example, the GES-DISC Interactive Online Visualization and Analysis Infrastructure (GIOVANNI) (http://giovanni.gsfc.nasa.gov/) TRMM Online Visualization and Analysis System (TOVAS) (http://disc.sci.gsfc.nasa.gov/precipitation/tovas) allows users to analyze and visualize most of the TRMM precipitation products without downloading any data and software (Acker and Leptoukh 2007; Berrick et al. 2009; Liu et al. 2007). GIOVANNI TOVAS has opened a new avenue for easy and friendly use of global remote sensing data, especially for those who are not familiar with NASA hierarchical data format (HDF) as well as other data formats, such as users from the socioeconomic community (Liu et al. 2007, 2010, 2011, 2012; Vicente and the GES DAAC Hydrology Data Support Team, 2007). Other than GIOVANNI TOVAS, the GES DISC has developed other data services to support precipitation applications around the world.

With continuing efforts to improve precipitation products in both spatial and temporal resolutions, new and higher resolution products are being developed, such as the next generation of TMPA, to be released in 2014, the Integrated Multi-satellitE Retrievals for GPM (IMERG). The IMERG product suite (http://pmm.nasa.gov/sites/default/files/document_files/IMERG_ATBD_V2.0.doc) will be significantly improved compared to those in TMPA in terms of spatial and temporal resolutions (from 0.25° to 0.1° and from three hourly to half hourly). The IMERG products consist of: output products, *early satellites* (lag time, ~4 h), *late satellites* (lag time, ~18 h), and the final *satellite-gauge* (lag time, ~2 months), along with additional new input and intermediate files, creating a new environment for intersatellite and intersensor intercomparison research, which is not available at present. However, the IMERG data volume will grow from ~29 GB/year (TMPA 3B42) to ~9 TB/year without counting the input and intermediate files. This jump in data volumes poses a great challenge to online data service and tool development in timely delivering of data services (data subsetting, format conversion, etc.), data analysis, and visualization to users around the world because current software cannot handle such file size in data analysis and visualization (due to memory fault and time out in web services). Issues related to increasing heterogeneity (e.g., complexity in data structure, file formats) can create a roadblock for online data service development as well. Big data (i.e., Snijders et al. [2012]), usually refers to datasets with sizes beyond the ability of commonly used software tools to capture, curate, manage, and process the data within a tolerable elapsed time. Nonetheless, it is now the *big data* era for online precipitation data services.

In this chapter, we first introduce the existing precipitation products and services at the GES DISC in order to better understand associated data issues, then we discuss challenges and solutions associated with the next precipitation satellite mission. In addition, a prototype that has been developed to address some of these issues is presented. This chapter is organized as follows: Section 5.2 presents an overview of global precipitation products and data services at the GES DISC, Section 5.3 the big data challenges and solutions, Section 5.4 a prototype, and the last section the conclusion.

5.2 OVERVIEW OF GLOBAL PRECIPITATION PRODUCTS AND DATA SERVICES

5.2.1 TRMM BACKGROUND

Precipitation is a critical component of the Earth's hydrological cycle. Launched on November 27, 1997, the NASA TRMM is a joint US–Japan satellite mission to monitor tropical and subtropical (40°N–40°S) precipitation and to estimate its associated latent heating. The TRMM satellite provides the first detailed and comprehensive dataset of the 4D distribution of rainfall and latent heating over vastly under-sampled tropical and subtropical oceans and continents (Special Issue on TRMM 2000). There are five instruments onboard the TRMM satellite, and three of them are used for precipitation. Standard TRMM products from the visible and infrared scanner (VIRS), the TRMM microwave imager (TMI), and the precipitation radar (PR) are archived at and distributed from the NASA GES DISC. Data products from the Clouds and Earth's Radiant Energy System (CERES) and the lightning imaging sensor (LIS) are archived at the Atmospheric Science Data Center (ASDC) at the NASA Langley Research Center and the NASA Global Hydrology Resource Center (GHRC), respectively. Table 5.1 lists the descriptions of the three precipitation-related instruments.

In August 2001, the TRMM satellite was boosted from 350 to 402.5 km, to extend its lifespan by reducing the consumption rate of the fuel used to maintain its orbit altitude. As of today, TRMM is still in operation and continuously collecting data. Since 1997, more than 15 years of TRMM data have been collected.

5.2.2 TRMM PRODUCTS

TRMM data products archived at and distributed from the GES DISC (Liu et al. 2012; Vicente and the GES DAAC Hydrology Data Support Team, 2007) are organized as

TABLE 5.1
TRMM Precipitation-Related Instruments

Instrument Name	Description
VIRS	Five channels (0.63, 1.6, 3.75, 10.8, and 12 μm); spatial resolution, 2.2 km (preboost) and 2.4 km (postboost); swath width, 720 km (preboost) and 833 km (postboost)
TMI	Five frequencies (10.7, 19.4, 21.3, 37, 85.5 GHz); spatial resolution, 4.4 km (at 85.5 GHz, preboost) and 5.1 km (at 85.5 GHz, postboost); swath width, 760 km (preboost) and 878 km (postboost)
PR	13.8 GHz; spatial resolution, 4.3 km (preboost) and 5.0 km (postboost); swath width, 215 km (preboost) and 247 km (postboost)

the following three categories: (1) orbital products (also known as swath products), (2) gridded products, and (3) other TRMM-related products, consisting of TRMM ancillary products, ground-based instrument products, TRMM and ground observation subsets, and field experiment products. Table 5.2 lists raw and calibrated satellite swath data, as well as geophysical swath products derived from VIRS, TMI, PR, and combined TMI/PR, such as 2A12 TMI hydrometeor profiles, 2A23 radar rain characteristics, and 2B31 combined rainfall profile. Table 5.3 contains monthly gridded products from single or multiple instruments, spatially and temporally averaged, and a daily gridded product. For example, 3A46 provides global rain rate from SSM/I alone. The collection of these monthly products allows intercomparison to understand precipitation biases and uncertainty. Two multi-satellite products are the

TABLE 5.2
Standard TRMM Version Seven Orbital Data Products
(Time Coverage: 12/1997–Present)

Product Name	Description
1A01: VIRS raw data (VIRS)	Reconstructed, unprocessed VIRS (0.63, 1.6, 3.75, 10.8, and 12 μm) data
1A11: TMI raw data (TMI)	Reconstructed, unprocessed TMI (10.65, 19.35, 21, 37, and 85.5 GHz) data
1B01: Visible and infrared radiance (VIRS)	Calibrated VIRS (0.63, 1.6, 3.75, 10.8, and 12 μm) radiances at 2.4 km resolution over a 833 km swath
1B11: Microwave brightness temperature (TMI)	Calibrated TMI (10.65, 19.35, 21, 37, and 85.5 GHz) brightness temperatures at 5–45 km resolution over a 878 km swath
1B21: Radar power (PR)	Calibrated PR (13.8 GHz) power at 5 km horizontal and 250 m vertical resolutions over a 247 km swath
1C21: Radar reflectivity (PR)	Calibrated PR (13.8 GHz) reflectivity at 5 km horizontal and 250 m vertical resolutions over a 247 km swath
2A12: Hydrometeor profile (TMI)	TMI hydrometeor (cloud liquid water, prec. water, cloud ice, prec. ice) profiles in 28 layers at 5.1 km (at 85.5 GHz) horizontal resolution, along with latent heat and surface rain, over a 878 km swath
2A21: Radar surface cross section (PR)	PR (13.8 GHz) normalized surface cross section at 5 km horizontal resolution and path attenuation (in case of rain), over a 247 km swath
2A23: Radar rain characteristics (PR)	Rain type: storm, freezing, and bright band heights; from PR (13.8 GHz) at 5 km horizontal resolution over a 247 km swath
2A25: Radar rainfall rate and profile (PR)	PR (13.8 GHz) rain rate, reflectivity, and attenuation profiles, at 5 km horizontal and 250 m vertical resolutions, over a 247 km swath
2B31: Combined rainfall profile (PR, TMI)	Combined PR/TMI rain rate and path-integrated attenuation at 5 km horizontal and 250 m vertical resolutions, over a 247 km swath

TABLE 5.3

Standard TRMM Version Seven Gridded Data Products

Data Product	Description	Time Range
3A11: Monthly 5 × 5° oceanic rainfall	Rain rate, conditional rain rate, rain frequency, and freezing height for a latitude band from 40°N to 40°S, from TMI	1997–12 to present
3A12: Monthly 0.5 × 0.5° mean 2A12, profile, and surface rainfall	0.5 × 0.5° gridded monthly product comprising mean 2A12 data and calculated vertical hydrometeor profiles, as well as mean surface rainfall	1997–12 to present
3A25: Monthly 5 × 5° and 5 × 0.5° spaceborne radar rainfall	Total and conditional rain rate, radar reflectivity, path-integrated attenuation at 2, 4, 6, 10, 15 km for convective and stratiform rain; storm, freezing, and bright band heights and snow ice layer depth for a latitude band from 40°N to 40°S, from PR	1997–12 to present
3A26: Monthly 5 × 5° surface rain total	Rain rate probability distribution at surface; 2 and 4 km for a latitude band from 40°N to 40°S, from PR	1997–12 to present
3A46 (Version 2): Monthly 1 × 1° SSM/I rain	Global rain rate from SSM/I	1998–01 to 2009–09
3B31: Monthly 5 × 5° combined rainfall	Rain rate, cloud liquid water, rain water, cloud ice, graupels at 14 levels for a latitude band from 40°N to 40°S, from PR and TMI	1997–12 to present
3B42: 3-h 0.25 × 0.25° merged TRMM and other satellite estimates	Calibrated IR merged with TRMM and other satellite data	1998–01 to present
3B42 daily: Daily 0.25 × 0.25° merged TRMM and other satellite estimates	Daily TRMM and other rainfall estimates	1998–01 to present
3B43 : Monthly 0.25 × 0.25° merged TRMM and other sources estimates	Merged 3B-42 and rain gauge estimates	1998–01 to present

most popular: the three-hourly and monthly TMPA products (3B42, 3B43), because of their better temporal and spatial resolutions, and the daily product derived from 3B42 for those who do not want high temporal resolution products.

Table 5.4 lists other TRMM-related products. The National Oceanic and Atmospheric Administration (NOAA) Climate Prediction Center (CPC) globally merged (60°N–60°S), half-hourly, 4 km IR brightness temperature data (equivalent blackbody temperatures, merged from several geostationary satellites around the globe) are an important ancillary product not only for precipitation algorithm development but also for providing background information for TRMM and other meteorological event case studies. Data from ground-based instruments provide radar data

TABLE 5.4
Other TRMM-Related Products

Product	Description
Ancillary	Globally merged (60°N–60°S) pixel-resolution IR brightness temperature data (equivalent blackbody temps), merged from all available geostationary satellites (GOES-8/10, METEOSAT-7/5). Associated Satellite ID files are available via ftp. (2000–02 to present).
Ground-based instruments	Ground-based instrument (radar data products from 10 TRMM project ground stations.
Subsets	Ground validation CSI data: The single VOS when the satellite nadir is within a specified distance from a ground validation or experiment site or a gridded field associated with the VOS that is coincident with a satellite overpass.
	Gridded subsets of orbital data products derived from VIRS, TMI, and PR.
	Satellite CSI data: Collection of instrument scan data when TRMM satellite nadir is within a specified distance from a ground validation or experiment site.
Field experiments	Provide ground truth for use in algorithm development for TRMM satellite measurements. The data archived at GES DISC include KWAJEX, LBA, SGP97, SGP99, SCSMEX, TEFLUNA, TEFLUNB, TOGA COARE, and TRMM LBA.

products from 10 TRMM project-affiliated ground stations in the tropical and subtropical regions. Table 5.4 describes subsets from (1) ground validation coincidence subsetted intermediate data (CSI), consisting of a single volume scan (VOS) when the satellite nadir is within a specified distance from a ground validation site or a gridded field associated with a VOS that is coincident with a satellite overpass; (2) gridded subsets of orbital data products derived from VIRS, TMI, and PR; and (3) collection of instrument scan data when the TRMM satellite nadir is within a specified distance from a ground validation or experiment site. These value-added subsets facilitate TRMM ground validation and other research activities, because users do not need to download the entire original orbital data and perform the subsetting task themselves.

The TRMM field campaign program was designed to provide independent ground truth for use in algorithm development for TRMM satellite measurements. TRMM field campaigns employ ground-based radars, rain gauge networks, and aircraft measurements from NASA DC-8 and ER2, with instrumentation similar to TMI and PR. TRMM field campaigns consist of TExas-FLorida UNderflight (TEFLUN A) and TEFLUN B, Large-Scale Biosphere-Atmosphere Experiment in Amazonia (TRMM-LBA), Kwajalein Experiment (KWAJEX), South China Sea Monsoon Experiment (SCSMEX), Convection And Moisture EXperiment (CAMEX), and Tropical Ocean Global Atmospheres/Coupled Ocean Atmosphere Response Experiment (TOGA COARE).

5.2.3 TRMM DATA SERVICES

Providing TRMM data services is very important for expediting research and applications and maximizing the societal benefits from TRMM (Liu et al. 2012). For many users, using remote sensing products can be a daunting task due to a number of problems, such as data format conversion, large data volume, and lack of software, among other problems. Useful data services can reduce data processing time and, thus, increase the time spent on scientific investigations and applications. Users generally are more likely to evaluate and use TRMM products if user-friendly data services are provided. Since TRMM was launched, several data services (Table 5.5) have been developed and/or applied at the GES DISC. Mirador is designed to facilitate data searching, accessing, and downloading. Mirador consists of a search and access web interface developed in response to the search habits of data users. It has a drastically simplified, clean interface and employs the Google mini appliance for metadata keyword searches. Other features include quick response, data file hit estimator, gazetteer (geographic search by feature name capability), and an interactive

TABLE 5.5
TRMM Data Services

Service	Description
Mirador	Mirador (from Spanish, *a place providing a wide view*) is a Google-based data archive search interface that allows searching, browsing, subsetting, format conversion, and ordering of Earth science data at NASA GES DISC.
TOVAS	A member of the GIOVANNI, which is the underlying infrastructure for a growing family of web interfaces that allows users to analyze gridded data interactively online without having to download any data.
TRMM read software	Read in a TRMM HDF data file and write out user-selected SDS arrays and Vdata tables as separate flat binary files.
Simple Subset Wizard	A simple spatial subset tool that allows spatial subsetting; outputs are in NetCDF.
REVERB	Refine your granule search with the NASA-developed Earth Observing System (EOS) Clearinghouse (ECHO) next-generation Earth science discovery tool.
GrADS Data Server	Stable, secure data server that provides subsetting and analysis services across the Internet. The core of GDS is OPeNDAP (also known as DODS), a software framework used for data networking that makes local data accessible to remote locations.
OPeNDAP	The OPeNDAP provides remote access to individual variables within datasets in a form usable by many tools, such as IDV, McIDAS-V, Panoply, Ferret, and GrADS.
OGC WMS	The OGC WMS provides map depictions over the network via a standard protocol, enabling clients to build customized maps based on data coming from a variety of distributed sources.

shopping cart. Value-added services include several data format conversions and spatial subsetting for a number of popular products.

To enable scientific exploration of Earth science data products without performing complicated and often time-consuming data processing steps (data downloading, data processing, etc.), the GES DISC has developed the GIOVANNI, based on user support experience and in consultation with members of the user community. The latter has requested quick data search, subset, analysis, display, and download capabilities. As a member of the GIOVANNI family, TOVAS allows accessing precipitation data products without downloading data or software.

Users of TRMM products can benefit from several other data services listed in Table 5.5. The TRMM read software developed at the GES DISC can read in all TRMM standard products and write out user-selected parameter arrays and other data in flat binary or ASCII files. The Simple Subset Wizard tool allows spatial subsetting and provides outputs in NetCDF. REVERB is a tool that allows keyword, spatial, and temporal search. The Grid Analysis and Display System (GrADS) (http://grads.iges.org/grads/grads.html) Data Server (GDS) (formerly known as GrADS-DODS Server) is a stable, secure data server that provides subsetting and analysis services across the Internet and provides a convenient way for GrADS users to access TRMM data. The core of GDS is the Open-Source Project for a Network Data Access Protocol (OPeNDAP) (also known as DODS), which provides remote access to individual variables within datasets in a form usable by many tools, such as Interactive Data Viewer (IDV), McIDAS-V, Panoply, Ferret, and GrADS. The Open Geospatial Consortium (OGC) Web Map Service (WMS) provides map depictions over the network via a standard protocol and enables clients to build customized maps with data coming from different networks.

Supporting applications is one of the missions at the GES DISC. For example, the US Department of Agriculture's Foreign Agricultural Service (USDA-FAS), in collaboration with the NASA GES DISC, is routinely using TMPA data to monitor precipitation and crop production around the world. This project is unique, being the first of its kind to utilize near real-time global satellite precipitation data in an operational manner (Liu et al. 2010). Satellite precipitation products are produced by the GES DISC via a semiautomated process and made publicly accessible from the USDA-FAS' Crop Explorer Web site (Liu et al. 2010). Monitoring precipitation for agriculturally important areas around the world greatly assists the USDA-FAS to quickly locate regional weather events, as well as help improve crop production estimates. Table 5.6 lists sample precipitation applications supported by the GES DISC.

5.2.4 GLOBAL PRECIPITATION MEASUREMENT MISSION

The GPM mission, initiated by NASA and the Japan Aerospace Exploration Agency (JAXA), is built upon the success of TRMM. The GPM mission consists of an international network of satellites that provide the next-generation global observations of rain and snow (http://pmm.nasa.gov/GPM). The concept of GPM is based on the deployment of a *core* satellite carrying an advanced radar/radiometer system to measure precipitation from space and serve as a reference standard to unify precipitation measurements from a constellation of research and operational satellites (http://pmm.nasa.gov/GPM). The GPM Core Observatory is scheduled for launch in early 2014. With improved

TABLE 5.6

Examples of TRMM Data Applications

Country of User	Application Description
International	International Society for Agriculture Meteorology
Afghanistan	Afghanistan Wildlife Conservation Society—conservation of wildlife efforts
Africa, Southeast Asia	MicroEnsure—develop a crop insurance product for developing countries
Australia	Environmental causes of diabetes. Rainfall as an effect on crop moisture and toxins
Australia	Analyze historical rainfall data in Southeast Asia for climate regionalization and stream flow
Australia	Study historical rainfall data over Thailand
France	Mean monthly rainfall in Northern Uganda
France	Daily historical rainfall in the Dominican Republic for crop insurance
Germany	Regionalization of marginality of agricultural land in West Africa
Iran	Study of ENSO impacts on Southeast Iran climate
Italy	Information Technology for Humanitarian Assistance (ITHACA)—early warning system of heavy rainfall
United Kingdom	Simulation of precipitation events in Florida
United Nations	The World Food Programme. Used rainfall for monitoring maize yield potential in East Timor and Indonesia
United States	Analyzed tropical rainfall and dust transport
United States	A study to find the correlation between rainfall and drought conditions in Afghanistan
United States	Correlating US monthly rainfall with Pacific and Atlantic sea surface temperatures
United States	A study of the monsoons over Southeast Asia and Ethiopia
United States	Used TRMM data for an early warning system for mosquito-borne diseases
United States	Used TRMM data in different regions to determine correlation between rainfall and malaria
United States	Analyzed precipitation over Upper Oconee Basin in Georgia
United States	Used TRMM monthly rainfall for the Amazon Basin, together with tree plot data to derive climate correlations of species diversity, biomass, and community structure in rain forests

measurements of precipitation globally, GPM will help advance our understanding of the Earth's water and energy cycle, improve forecasting of extreme events that cause natural hazards and disasters, and extend current capabilities in using accurate and timely information of precipitation to directly benefit society.

The first spaceborne Ku/Ka-band dual-frequency precipitation radar (DPR) and a multichannel GPM microwave imager (GMI) will be carried by the GPM Core Observatory (http://pmm.nasa.gov/GPM). Similar to the TRMM PR, the DPR instrument will provide 3D measurements of precipitation structure. However, the DPR instrument will add a new Ku-band precipitation radar (KuPR) operating at 13.6 GHz (http://pmm.nasa.gov/GPM). There are several new features with the DPR. Firstly, the DPR is more sensitive to light rain rates and snowfall by comparing to the TRMM PR. Secondly, simultaneous measurements by the overlapping of Ka/Ku-bands of the DPR provide new information on particle drop size distributions over moderate precipitation intensities (http://pmm.nasa.gov/GPM). Lastly, the improved sensitive measurement capability for microphysical properties can complement cloud and aerosol observations, providing further insights into how precipitation processes may be affected by human activities (http://pmm.nasa.gov/GPM).

The GMI instrument is a conical-scanning multichannel microwave radiometer covering a swath of 550 miles (885 km) with 13 channels ranging in frequency from 10 to 183 GHz (http://pmm.nasa.gov/GPM). These GMI frequencies have been optimized over the past two decades to retrieve heavy, moderate, and light precipitation using the polarization difference at each channel as an indicator of the optical thickness and water content (http://pmm.nasa.gov/GPM). In the GPM era, there will be 8–10 functioning satellites across the different phases of its anticipated mission lifetime.

Over the past decade, algorithms that blend data from other satellites have been developed to address limited spatial and temperature coverage issues in TRMM. In particular, the IMERG product suite (http://pmm.nasa.gov/sites/default/files/document_files/IMERG_ATBD_V2.0.doc) will be significantly improved comparing those in TMPA in terms of spatial and temporal resolutions (from 0.25° to 0.1° and from three hourly to half hourly). It contains three output products, *early satellites* (lag time, ~4 h), *late satellites* (lag time, ~18 h), and the final *satellite-gauge* (lag time, ~2 months), along with additional new input and intermediate files, creating a new environment for intersatellite and intersensor intercomparison research, which is not available at present. However, the IMERG data volume will grow from ~29 GB/year (TMPA 3B42) to ~9 TB/year without counting the input and intermediate files.

With additional new instruments, new frequencies, improved blended algorithms, and ground validation measurements, it is expected that the data volume during the GPM era will grow significantly, ~20 times of the TRMM archive.

5.3 BIG DATA CHALLENGES AND SOLUTIONS

5.3.1 BIG DATA CHALLENGES

According to Snijders et al. (2012), *big data* usually refers to datasets with sizes beyond the ability of commonly used software tools to capture, curate, manage, and process the data within a tolerable elapsed time. Based on the description in Section 5.2, TRMM and GPM precipitation products meet the definition of *big data* due to the following challenges. The improvement in both spatial and temporal resolutions during the GPM era (i.e., IMERG) poses a challenge in data analysis and visualization due to the jump in data volume. For example, the IMERG data volume will grow from ~29 GB/year (TMPA 3B42) to ~9 TB/year. The increasing file size can cause problems in existing analysis and visualization software in timely delivery of analysis and visualization for online tools (i.e., GIOVANNI) as presented in Section 5.2.

The large volume can easily cause a time out in web services for online data analysis due to either software crash or time needed for data processing. The increasing volume of data will affect other existing data services, including data subsetting and data format conversion, among other things, all of which are needed to support online data services.

For TRMM level 2 orbital products in Section 5.2, it has been a challenge since TRMM, and no online system exists by far to analyze and visualize TRMM orbital data due to a large memory requirement and complex data structures. For the time being, visualization of TRMM orbital data is done on a Windows environment, such as the Orbital Viewer (http://pps.gsfc.nasa.gov/THOR/release.html). During the GPM era, the number of satellites will increase from current TRMM to 8–10 other satellites, involving more instruments and more data sources available for analysis and visualization. For example, there will be up to 12 microwave radiometers in the GPM era (http://pmm.nasa.gov/sites/default/files/document_files/GPROF_ATBD_1Dec2010.pdf), and it is a challenge to make these heterogeneous datasets available for online services.

Likewise, the heterogeneous data issue also exists in data from TRMM field campaigns that employ ground-based radars, rain gauge networks, and aircraft measurements from NASA DC-8 and ER2, with instrumentation similar to TMI and PR. Each measurement has its own data format and data structure, making data verification between TRMM and the field campaigns difficult and time consuming. This issue will exist in the GPM era as well. The same challenge applies to other TRMM and GPM ancillary data as well.

Another challenge is to provide long time series data services to communities such as the hydrology community (Teng et al. 2012). In hydrology, the Earth surface features are expressed as discrete spatial objects such as watersheds, river reaches, and point observation sites, and time-varying data are contained in time series associated with these spatial objects. Longtime histories of data may be associated with a single point or feature in space. Most remote sensing precipitation products are expressed as continuous spatial fields, with data sequenced in time from one data file to the next. Hydrology tends to be narrow in space and deep in time, which poses a challenge during the GPM era. For example, to generate a 1-year time series, one needs to pull all the $0.1°$, half-hourly IMERG product, which can be time consuming and not suitable for online data services due to the large volume of data.

For data applications, there are issues associated with the current plan for GPM products. An example is that there will be no daily and 10-day products available in the standard products, and both products are important for agriculture management activities (Liu et al. 2010). To generate these products on the fly through web services can be a time-consuming task due to issues such as file size, as mentioned earlier.

In short, the earlier challenges in developing online data services can be summarized as file size, data structure, data format, data integration with heterogeneous products, long time series, etc. Traditional information was derived from analysis of a single large set of related data. In the modern era, information is derived from multiple sources with approximately the same total amount of data, requiring new solutions to solve these problems. Nonetheless, the *big data* challenges exist in the online precipitation data services at the GES DISC.

5.3.2 SOLUTIONS

Solutions to the big data challenges can be classified as three categories: (1) developing new software, (2) developing algorithms that maximize the use of existing software, and (3) the combination of both 1 and 2. At the GES DISC, (1) is difficult due to budget constrain. Because (2) can not totally satisfy the need for processing new products from the GPM mission due to new data formats, data structures, data volumes, etc., (3) is being considered with an emphasis on the reuse of existing software. We describe this approach in the succeeding text in this section.

For the increasing data file issue during the GPM era, one solution is to break up a high-resolution product into several coarse spatial and temporal resolution products. This is not only for fast online visualization and analysis but also for other data services for users who do not want precipitation products at the original resolution in their research and applications. For example, some TRMM users do not want to download the three hourly product (3B42), and its derivation daily is sufficient for their studies. The GES DISC developed a new and nonstandard daily product to satisfy this need. For less data processing during data analysis and visualization, users who look for a rain event can first analyze the daily data and use the three hourly data for additional details, if needed. Likewise, daily products can be derived from the IMERG product suite. Because the temporal resolution of IMERG is half hourly, other products, such as hourly, three hourly, six hourly, daily, five day (pentad), and ten day (dekad), can be developed for different needs ranging from research, modeling, and applications (Liu et al. 2010, 2012). The increasing spatial resolution in IMERG also requires additional processing time for data analysis and visualization, and deriving products in different spatial resolutions can improve data processing time because not all users want the original spatial resolution of IMERG for various reasons. A prototype in Section 5.4 will describe this approach with examples.

Online visualization and analysis of level 2 swath data can be a challenge as mentioned earlier. Different algorithms use different data structures to best describe the contents in their products (http://pps.gsfc.nasa.gov/Documents/filespec.TRMM.V7.pdf; ftp://pps.gsfc.nasa.gov/pub/GPMfilespec/filespec.GPM.V1.pdf). Special software often needs to be developed for each product (e.g., Ostrenga et al. 2010) so the output can be in a uniform format (i.e., NetCDF) with which analysis and visualization can be easily done using off-the-shelf software. The YoTC project (Ostrenga et al. 2010) mentioned in Section 5.1 is an example that uses different software to read TRMM TMI and PR products and extract profiles at the point of interest (http://disc.sci.gsfc.nasa.gov/YOTC/yotc_gs). Likewise, this approach can be applied to field campaign data products as well.

Providing analysis and visualization for long time series data can be achieved through data rods (Gallaher and Grant 2012; Teng et al. 2012). Teng et al. (2012) proposed two general solutions for the data rod approach: (1) retrieve multiple time series for short time periods and stitch the multiple time series into desired single long time series and (2) reprocess (parameter and spatial subsetting) and archive data as one-time cost approach. The resultant time series files would be geospatially searchable and could be optimally accessed and retrieved by any user at any time (Teng et al. 2012). One drawback for the data rod approach is that there are a lot of files to be generated and maintained. For example, for IMERG, the number of files can be as many as $1300 \times 3600 = 4,680,000$ if one time series is saved for each grid point. To reduce the number of files, several grids can be grouped together to form clusters. The first approach is possible for parallel processing.

Different data formats can create an obstacle for analysis and visualization such as scatter plot and intercomparison, due to the need for conversion tools. Currently, GIOVANNI (Acker and Leptoukh 2007; Berrick et al. 2009) is using this approach to integrate products in different formats. GIOVANNI first fetches data from different data services, that is, ftp, OPeNDAP, and GDS, and then converts the fetched data into its internal format, HDF (Berrick et al. 2009). Some analytical functions (scatter plot, intercomparison, etc.) require the same grid structure for both gridded products. For gridded products in different spatial resolutions, a regridding software is needed to convert the grid structure of the high-resolution product to match that of the low resolution, or vice versa, though the former is used more frequently. For nongridded orbital products, it is more difficult because data locations from different sensors seldom match each other, and more complicated software is needed comparing to gridded products.

For the application challenge, the large volume IMERG products can be used to derive products in different spatial and temporal resolutions, as mentioned earlier. This approach can satisfy current application needs at the GES DISC. For example, the USDA-FAS' Crop Explorer can be supported with the IMERG derived 10-day product.

To test the solutions mentioned here, we have developed a prototype to analyze and visualize a TRMM ancillary product, the 4 km global merged IR brightness temperature product (Janowiak et al. 2001; Liu et al. 2011) since the IMERG product suite is currently not available. Both the global merged IR product and the IMERG share some similarities, such as, both are half hourly and gridded, although the former has a higher spatial resolution (4 km).

5.4 A PROTOTYPE

5.4.1 Data

Since the Nimbus satellite program (Bluestein 1993; Kidder and von der Harr 1995), IR (~ 11 μm channel) images recorded by satellite sensors have been widely used in public TV weather forecast, research, and classroom textbooks as an illustration of weather system movement and evolution. For example, if *IR image* is searched in the advanced search of the American Meteorological Society (AMS) online journal, it returns more than 5000 different article titles. Unlike visible images, IR images can reveal cloud features without sunlight illumination; therefore, IR images can be used to monitor weather phenomena day and night. With geostationary satellites deployed around the globe, it is possible to monitor global weather events 24/7 at a temporal resolution that polar orbiting satellites cannot achieve at the present time (Janowiak et al. 2001). When IR products from different geostationary satellites are merged to form a single product, also known as a merged product, it allows observing weather at a global scale simultaneously

(Janowiak et al. 2001). Its half-hourly temporal resolution also makes it an ideal ancillary dataset for supporting other satellite missions, such as, TRMM (Special Issue on TRMM 2000), and multisensor and multisatellite research by providing additional information about weather system evolution.

Satellite IR images have a wide range of applications as presented in research articles and textbooks (e.g., Bluestein 1993; Janowiak et al. 2001; Kidder and von der Harr 1995). Trained analysts use IR images to identify meteorological events with cloud signatures, such as, cloud types, fronts, mesoscale convective systems (MCS), and tropical storms, as described in meteorological textbooks (e.g., Bluestein 1993; Kidder and von der Harr 1995). With a series of consecutive images or animation, they can further track development and movement of weather systems. Researchers can use archived IR imagery to conduct case studies of weather events around the globe. Students can use IR imagery in classroom projects to assist in learning weather systems, satellite remote sensing, algorithm retrieval methods, and more. Nevertheless, satellite IR imagery is important for weather forecast, research, and education.

The GES DISC is home to TRMM data archive (Liu et al. 2012; Vicente and the GES DAAC Hydrology Data Support Team, 2007). It also houses an archive of the global merged IR brightness temperature product known as the NCEP/CPC 4 km, half-hourly global (60°N–60°S or 9896 longitude grids × 3298 latitude grids) IR dataset (Janowiak et al. 2001; Liu et al. 2012), which serves as one of TRMM's ancillary datasets. The global merged IR product consists of global IR brightness temperature data (equivalent blackbody temperatures) merged from operational geostationary satellites (US GOES east 75°W and west 135°W, European Meteosat 0°E and 57°E, and Japanese GMS/MTSAT 135°E). The GES DISC has collected more than 13 years of these data from February 2000 onward, creating an archive that is likely of particular importance to researchers, students, and others who are interested in the research and education of past weather events as many existing web services only focus on providing images for current weather events.

While the merged IR dataset can be downloaded directly via ftp, its large volume poses a challenge for many users. A single file occupies ~70 MB disk space and there is a total of ~115, 165 files (~7.7 TB, as of March 2013) for the past 13 years. To generate a single image, users need to go through the following steps:

- Need a computer equipped with adequate software for scientific data processing and visualization.
- Obtain the data.
- Study the data product document.
- Obtain a software for data processing or subsetting if available; write their own otherwise.
- Use an existing software or write their own to generate the imagery.

All of these steps could be a burden to many users, especially nonprofessional users and those in developing countries. Because significant investments in equipment and time are needed in these steps, the ability to explore data anytime anywhere could be limited. Many existing web services only focus on providing images for current weather events. Some provide historical pregenerated images that do not allow interactive imagery subsetting, display, and customized analysis. As information technology develops, it is possible to develop an online system that performs the tasks listed previously, allowing scientific users to focus more on scientific data analysis and discovery.

5.4.2 System Description

Online visualization and analysis of the merged IR product are a part of the Hurricane Data Analysis Tool (HDAT) (http://disc.gsfc.nasa.gov/HDAT, Liu et al. 2011). The HDAT's architecture is based on the GIOVANNI TOVAS (Acker and Leptoukh 2007; Berrick et al. 2009; Liu et al. 2007, 2009, 2010), which primarily consists of TRMM products and has been in operation since March 2000 to support various research, application, and education activities (Liu et al. 2007, 2010). The principle design goal (Acker and Leptoukh 2007; Berrick et al. 2009; Liu et al. 2007, 2009, 2010;) for GIOVANNI was to provide a quick and simple interactive means for science data users to study various geophysical phenomena by trying various combinations of parameters measured by different instruments, arrive at a conclusion, and then generate graphs suitable for publication (examples in, http://disc.sci.gsfc.nasa.gov/giovanni/additional/publications). Alternatively, GIOVANNI would provide means to ask relevant *what if* questions and get back answers that would stimulate further investigations. This would all be done without having to download and preprocess large amounts of data (Acker and Leptoukh 2007; Berrick et al. 2009; Liu et al. 2007, 2009, 2010).

Figure 5.1 shows a schematic system diagram of HDAT, consisting of HTML and Common Gateway Interface (CGI) scripts written in Perl, GrADS scripts, and scripts for image processing. In addition, there is an interactive image map through which a user can select a bounding box area to define an area of interest (Figure 5.2). GrADS was chosen for its widespread use for providing easy access, manipulation, and visualization of meteorological observations and model data. It supports a variety of data

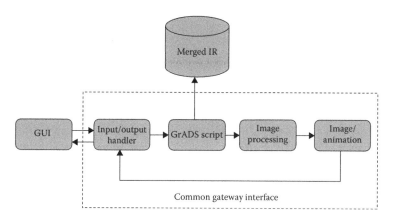

FIGURE 5.1   A schematic system diagram for the HDAT.

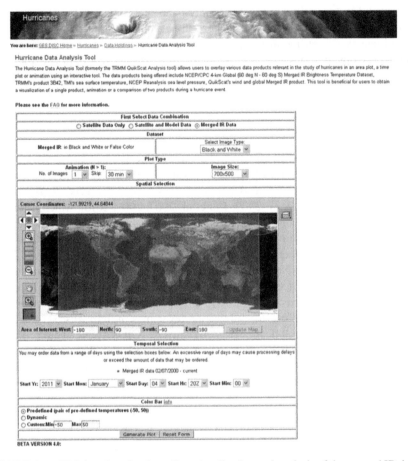

FIGURE 5.2 Web interface for the online visualization and analysis of the merged IR dataset as a part of HDAT. Basic functions include selection of an area of interest, single image or animation, a time skip capability for different temporal resolutions, and image size selection.

formats such as binary, GRIB, NetCDF, HDF, and HDF-EOS. Since the merged IR data is a gridded dataset (Janowiak et al. 2001), it can be read by GrADS; therefore, GrADS can be used directly as a backend analytical engine. The use of GrADS can significantly reduce development time because there is no need to rewrite software. Because GrADS is primarily for analyzing model and gridded data, additional scripts for image processing (Figure 5.1) are needed to generate black and white images, remove noises and missing data, enhance images, etc. (Campbell 2008; Russ 2007; Schowengerdt 2007). Currently, a script for generating black/white and false color imagery and animation has been implemented.

Handling a large original or raw data archive for an online analysis system is not an easy task. Existing systems primarily use pregenerated smaller image subsets or tiles and merge them when necessary to form a larger image (e.g., Google Map). If a system is designed for displaying images only, this method can work very well and efficiently; however, if a system allows users conducting online analysis and running algorithms on the fly, the ability for the system to access the original data is a must. Since the current system uses only the original dataset, it has a lot of potential to include additional analytical functions, such as image processing (Campbell 2008; Russ 2007; Schowengerdt 2007).

If the merged IR product at the full resolution (4 km) is directly used by GrADS, it can cause memory fault when an image that covers the entire data domain is displayed. Also, it is a time-consuming process and a waste of resources to use the original resolution data if the output image size is not large enough to display all the pixels or at least most of them. An efficient way to overcome this issue is to develop datasets at different resolutions. Based on the output image size, the system can be designed to match the nearest resolution dataset in terms of pixels and grids. For example, if a user selects a 3200 × 3200 km bounding area from GUI and the output image size is 400 × 400 pixels, an 8 km merged IR dataset is sufficient because only the image with an 800 × 800 pixel size or larger can use all the 4 km pixels. If there is no 8 km dataset, the system needs to read in all 9896 × 3298 grids, three times the amount of the pixels than the 8 km dataset. In reality, it is more complicated than this due to nonsquare shape, disk space limit, etc. For HDAT, we only generate a 32 km dataset using a box average method, and the threshold for switching dataset is a 4000 × 4000 (~40° × 40°) km area after balancing various factors. The choice for one dataset at the reduced resolution can sometimes cause grainy pixels for some selected areas larger than the threshold.

Through the HDAT web interface in Figure 5.2, a user can select the merged IR dataset and then choose a geographical region of interest, temporal extent, time step, and type of image to be produced. The output is displayed in a new window (Figure 5.3). Basic functions are listed in Table 5.7. Users can save an animation as a file (animated gif) and import it into other presentation software, such as Microsoft PowerPoint. The current capabilities allow users to generate up to 60 images in an animation with a time skip ranging from 30 min to 24 h, and the image size varies from 700 × 500 pixels up to 1400 × 1080 pixels.

5.4.3 EXAMPLES

Figure 5.3 is a large-scale image at 01 Z on October 3, 2009 showing the super typhoons Parma (western storm over the northern Philippines) and Melor (eastern storm approaching the northern Marianas). A close up of Parma is shown in continuous (Figure 5.3b) and discrete (Figure 5.3c, in degrees Celsius) grayscales, respectively. With HDAT, an animation can be generated on the fly (not shown) to study the interaction of two nearby typhoons or the Fujiwhara effect or Fujiwara interaction (Fujiwhara 1921).

Figure 5.4 shows that users can also retrieve images of weather events. For example, Figure 5.4a shows a strong cold front passage in the Mid-Atlantic region, and Figure 5.4b shows a record-breaking rainfall event in Mumbai, India, on July 26, 2005. Their animations can be also retrieved from HDAT to show the system evolution in every 30 min in either black and white or false color.

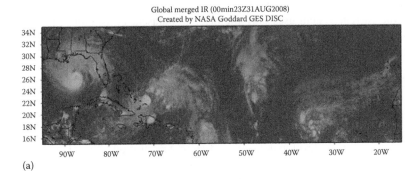

(a)

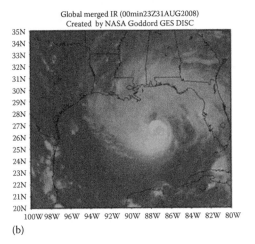

(b)

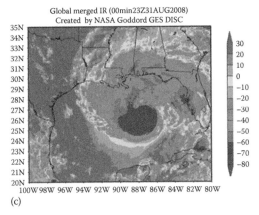

(c)

FIGURE 5.3 A large-scale image (a) at 01 Z on October 3, 2009, showing super typhoons Parma (western storm over the northern Philippines) and Melor (eastern storm approaching the northern Marianas). (b) and (c): A close-up of Parma in continuous and discrete (in degrees Celsius) grayscales, respectively.

TABLE 5.7

Functions in HDAT

Functions

Selection of area of interest

Image type (black/white, false color)

Single image or animation

Time increment (30 min–24 h)

Image size selection

Color bar for false color images

Minimum and maximum adjustments for color bar

Single image or animation can be saved as GIF or
   animated GIF

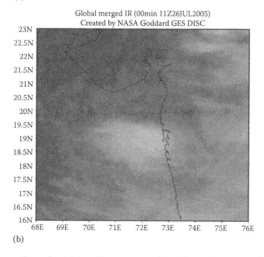

(a)

(b)

FIGURE 5.4 Examples of past weather events: (a) a frontal passage in the Mid-Atlantic region; (b) a record-breaking rainfall event in Mumbai, India.

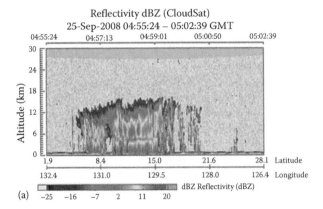

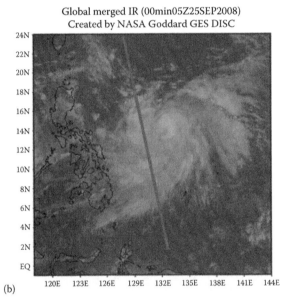

FIGURE 5.5 Example of using the merged IR product as ancillary information to support other satellite missions: (a) a CloudSat reflectivity snapshot for typhoon Jangmi; (b) the merged IR images at the closest time with an overlay of the CloudSat track (straight line).

The tool can be used to provide ancillary information to interpret measurements from other satellite missions. Figure 5.5 is an example of using the merged IR product as ancillary information to support other satellite missions. Figure 5.5a is a snapshot from CloudSat (Stephens et al. 2002) for typhoon Jangmi, and Figure 5.5b shows the merged IR image with an overlay of the CloudSat orbital track (straight line). Users can use HDAT to study the time evolution of Jangmi as well.

5.5   CONCLUSIONS

NASA satellite-derived global precipitation and their ancillary products are heterogeneous with different formats and complex data structures, often requiring custom software for data processing, analysis, and visualization. During the GPM era, more satellite products with improved spatial and temporal resolutions will be added into the data archive. File sizes will significantly increase along with more data formats and complicated data structures. Traditional data processing software will have difficulties to timely deliver online-based data services, in particular, online visualization and analysis. All these create *big data* challenges for data service providers. The challenging issues to be faced include (1) increasing product file sizes; (2) heterogeneous data structures in standard orbital and ancillary products; (3) different data formats and structures between standard, field campaign, and ancillary products; (4) delivering on-the-fly data services for long time series from products with high temporal resolution (i.e., half hourly); and (5) application requirements.

We have discussed several approaches at the GES DISC to these challenging issues. High-resolution products can be repackaged or regridded into coarse products in both space and time to satisfy needs of both online services and users. Heterogeneous products can be processed to obtain similar data structures for online services. Likewise, products with different formats can be reformatted, which can be used by existing software. Currently, GIOVANNI is using this approach to integrate products in different formats. Data rods can be used to provide on-the-fly long time series from high-resolution products. The aforementioned approaches can also be applied to applications.

As preparing for the release of the next-generation IMERG product suite, a prototype for online visualization and analysis of the global merged IR product has been developed at the GES DISC for testing the earlier approaches. Users can use it to generate customized merged IR images and animations for a period from February 2000 onward without downloading data and software. Despite a grainy pixel issue in some large area output images, the tool allows users to analyze the merged IR imagery in black/white or false color with some degrees of customization (i.e., color bar, image size). Further enhancements include adding additional datasets at reduced resolutions to reduce grainy pixels for all areas selected by users and more functions for image processing. However, the issue of how datasets at different spatial resolutions affect data analysis results (i.e., area-averaged time series) still remains and more studies are needed in the future. Nonetheless, HDAT can serve a prototype for the future IMERG product suite, and the HDAT approach can be integrated into the GES DISC GIOVANNI system.

ACKNOWLEDGMENT

This work was supported by the NASA GES DISC.

REFERENCES

Acker, J. G. and G. Leptoukh, 2007, Online analysis enhances use of NASA earth science data. *Eos. Trans. Amer. Geophys. Union*, 88(2), 14–17.

Anthes, R. et al., 2007, *Earth Science and Applications from Space: National Imperative for the Next Decade and Beyond*, National Academies Press, Washington, DC, pp. 61–78.

Aonashi, K., J. Awaka, M. Hirose, T. Kozu, T. Kubota, G. Liu, S. Shige et al., 2009, GSMaP passive, microwave precipitation retrieval algorithm: Algorithm description and validation. *J. Meteorol. Soc. Japan*, 87A, 119–136.

Behrangi, A., K.-L. Hsu, B. Imam, S. Sorooshian, G. J. Huffman, and R. J. Kuligowski, 2009, PERSIANN-MSA: A precipitation estimation method from satellite-based multispectral analysis. *J. Hydrometeorol.*, 10, 1414–1429. doi: http://dx.doi.org/10.1175/2009JHM1139.1.

Berrick, S. W., G. Leptoukh, J. D. Farley, and H. Rui, 2009, Giovanni: A web service workflow-based data visualization and analysis system. *IEEE Trans. Geosci. Remote Sens.*, 47(1), 106–113.

Bluestein, H., 1993, *Synoptic-Dynamic Meteorology in Midlatitudes: Observations and Theory of Weather Systems*, Vol. 2, Oxford University Press, New York, p. 594.

Campbell, J. B., 2008, *Introduction to Remote Sensing*, 4th edn., The Guilford Press, New York, p. 626.

Chiu, L. S., Z. Liu, J. Vongsaard, S. Morain, A. Budge, P. Neville, and C. Bales, 2006, Comparison of TRMM and water district rain rates over New Mexico. *Adv. Atmos. Sci.*, 23(1), 1–13.

Chokngamwong, R. and L. S. Chiu, 2008, Thailand daily rainfall and comparison with TRMM products. *J. Hydrometeorol.*, 9(2), 256–266.

Fujiwhara, S., 1921, The natural tendency towards symmetry of motion and its application as a principle in meteorology. *Quart. J. Roy. Meteorol. Soc.*, 47(200), 287–293.

Gallaher, D. and G. Grant, 2012, Data rods: High speed, time-series analysis of massive cryospheric data sets using pure object databases, *Geoscience and Remote Sensing Symposium (IGRASS)*, Munich, Germany, July 22–27, 2012 [available: http://ieeexplore.ieee.org/xpl/articleDetails.jsp?reload=true&arnumber=6352413&contentType=Conference+Publications].

Habib, E., A. T. Haile, Y. Tian, and R. J. Joyce, 2012, Evaluation of the high-resolution CMORPH satellite rainfall product using dense rain gauge observations and radar-based estimates. *J. Hydrometeorol.*, 13, 1784–1798. doi: http://dx.doi.org/10.1175/JHM-D-12-017.1.

Hong, Y., D. Gochis, J.-T. Cheng, K.-L. Hsu, and S. Sorooshian, 2007, Evaluation of PERSIANN-CCS rainfall measurement using the NAME event rain gauge network. *J. Hydrometeorol.*, 8, 469–482. doi: http://dx.doi.org/10.1175/JHM574.1.

Huffman, G. J., R. F. Adler, P. Arkin, A. Chang, R. Ferraro, A. Gruber, J. Janowiak, A. McNab, B. Rudolph, and U. Schneider, 1997, The global precipitation climatology project (GPCP) combined precipitation dataset. *Bull. Am. Meteorol. Soc.*, 78, 5–20.

Huffman, G. J., R. F. Adler, D. T. Bolvin, G. Gu, E. J. Nelkin, K. P. Bowman, Y. Hong, E. F. Stocker, and D. B. Wolff, 2007, The TRMM multi-satellite precipitation analysis: Quasi-global, multi-year, combined-sensor precipitation estimates at fine scale. *J. Hydrometeorol.*, 8(1), 38–55.

Huffman, G. J., R. F. Adler, D. T. Bolvin, and E. J. Nelkin, 2010, The TRMM multi-satellite precipitation analysis (*TAMPA*). In *Satellite Rainfall Applications for Surface Hydrology*, F. Hossain and M. Gebremichael, eds. Springer Verlag, Dordrecht, the Netherlands, pp. 3–22.

Huffman, G. J., R. F. Adler, M. Morrissey, D. T. Bolvin, S. Curtis, R. Joyce, B. McGavock, and J. Susskind, 2001, Global precipitation at one-degree daily resolution from multi-satellite observations. *J. Hydrometeorol.*, 2(1), 36–50.

Huffman, G. J., R. F. Adler, B. Rudolph, U. Schneider, and P. Keehn, 1995, Global precipitation estimates based on a technique for combining satellite-based estimates, rain gauge analysis, and NWP model precipitation information. *J. Climate*, 8, 1284–1295.

Janowiak, J. E., R. J. Joyce, and Y. Yarosh, 2001, A real-time global half-hourly pixel-resolution infrared dataset and its applications. *Bull. Am. Meteorol. Soc.*, 82(3), 205–217.

Joyce, R. J., J. E. Janowiak, P. A. Arkin, and P. Xie, 2004, CMORPH: A method that produces global precipitation estimates from passive microwave and infrared data at high spatial and temporal resolution. *J. Hydrometeorol.*, 5, 487–503.

Kidder, S. Q. and T. H. von der Haar, 1995, *Satellite Meteorology—An Introduction*, Academic Press, San Diego, CA, p. 466.

Liu, Z., S. Kempler, W. Teng, H. Rui, L. S. Chiu, and L. Milich, 2010, Global agriculture information system and its applications, in *Other Applications, Advanced Geoinformation Science*, C. Yang, D. Wong, Q. Miao, and R. Yang, eds. CRC Press, Taylor & Francis, Boca Raton, FL, pp. 353–361.

Liu, Z., D. Ostrenga, and G. Leptoukh, 2011, Online visualization and analysis of global half-hourly pixel-resolution infrared dataset. *Bull. Am. Meteorol. Soc.*, 92(4), 429–432, doi: http://dx.doi.org/10.1175/2010BAMS2976.1.

Liu, Z., D. Ostrenga, W. Teng, and S. Kempler, 2012, Tropical rainfall measuring mission (TRMM) precipitation data and services for research and applications. *Bull. Am. Meteorol. Soc.*, 93(4), 1317–1325, doi: http://dx.doi.org/10.1175/BAMS-D-11-00152.1.

Liu, Z., H. Rui, W. Teng, L. Chiu, G. Leptoukh, and G. Vicente, 2007, Online visualization and analysis: A new avenue to use satellite data for weather, climate and interdisciplinary research and applications. Measuring precipitation from space—EURAINSAT and the future. *Adv. Glob. Change Res.*, 28, 549–558.

Liu, Z., H. Rui, W. L. Teng, L. S. Chiu, G. G. Leptoukh, and S. Kempler, 2009, Developing an online information system prototype for global satellite precipitation algorithm validation and intercomparison. *J. Appl. Meteorol. Climatol.-IPWG Special Issue*, 48(12), 2581–2589.

Mahrooghy, M., V. G. Anantharaj, N. H. Younan, J. Aanstoos, and K.-L. Hsu, 2012, On an enhanced PERSIANN-CCS algorithm for precipitation estimation. *J. Atmos. Oceanic Technol.*, 29, 922–932. doi: http://dx.doi.org/10.1175/JTECH-D-11-00146.1.

Moncrieff, M. W., D. E. Waliser, M. J. Miller, M. A. Shapiro, G. R. Asrar, and J. Caughey, 2012, Multiscale convective organization and the YOTC virtual global field campaign. *Bull. Am. Meteorol. Soc.*, 93, 1171–1187. doi: http://dx.doi.org/10.1175/BAMS-D-11-00233.1.

Ostrenga, D., G. Leptoukh, D. Waliser, Z. Liu, and A. Savtchenko, 2010, *NASA Giovanni Tool for Visualization and Analysis Support for the YOTC Program. AGU Fall Meeting*, San Francisco, CA.

Obasi, G. O. P., 1994, WMO's role in the international decade for natural disaster reduction. *Bull. Amer. Meteor. Soc.*, 75, 1655–1661. doi: http://dx.doi.org/10.1175/1520-0477(1994)075<1655:WRITID>2.0.CO;2

Petherick, A., 2012, Enumerating adaptation. *Nature Clim. Change*, 2, 228–229. doi: 10.1038/nclimate1472, Published online March 28, 2012.

Rozante, J. R., D. S. Moreira, L. G. G. de Goncalves, and D. A. Vila, 2010, Combining TRMM and surface observations of precipitation: Technique and validation over South America. *Weather and Forecasting*, 25(3), 885–894.

Russ, J. C., 2007, *The Image Processing Handbook*, 5th edn., CRC Press, Boca Raton, FL, p. 817.

Schowengerdt, R. A., 2007, *Remote Sensing Models and Methods for Image Processing*, 3rd edn., Academic Press, Burlington, MA, p. 515.

Snijders, C., Matzat, U., and Reips, U.-D., 2012, "Big Data": Big gaps of knowledge in the field of internet science. *Int. J. Internet Sci.*, 7, 1–5. (Available: http://www.ijis.net/ijis7_1/ijis7_1_editorial.html)

Sorooshian, S., K.-L. Hsu, X. Gao, H. V. Gupta, B. Imam, and D. Braithwaite, 2000, Evaluation of PERSIANN system satellite–based estimates of tropical rainfall. *Bull. Am. Meteorol. Soc.*, 81, 2035–2046. doi: http://dx.doi.org/10.1175/1520-0477(2000)081<2035:EOPS SE>2.3.CO;2

Special Issue on the Tropical Rainfall Measuring Mission (TRMM), Combined publication of the December 2000 Journal of Climate and Part 1 of the December 2000. *Journal of Applied Meteorology*, American Meteorological Society, Boston, MA.

Stephens, G. L., D. G. Vane, R. J. Boain, G. G. Mace, K. Sassen, Z. Wang, A. J. Illingworth et al., 2002, The CloudSat mission and the A-Train: A new dimension of space based observations of clouds and precipitation. *Bull. Am. Meteorol. Soc.*, 83, 1771–1790.

Teng, B., D. R. Maidment, B. Vollmer, C. Peters-Lidard, H. Rui, R. Strub, T. Whiteaker, D. Mocko, and D. Kirschbaum, 2012, Bridging the digital divide between discrete and continuous space-time array data to enhance accessibility to and usability of NASA Earth Sciences data for the hydrological community. *AGU Fall Meeting*, December 3–7, San Francisco, CA.

Tian, Y. and C. Peters-Lidard, 2010, A global map of uncertainties in satellite-based precipitation measurements. *Geophys. Res. Lett.*, 37(L24407), 1–6. (10.1029/2010GL046008).

Tian, Y., C. Peters-Lidard, and B. John, 2010, Real-time bias reduction for satellite-based precipitation estimates. *J. Hydrometeorol.*, 11, 1275–1285. (10.1175/2010JHN1246.1).

Vicente, G. A. and the GES DAAC Hydrology Data Support Team, 2007, Global satellite datasets: Data availability for scientists and operational users. Measuring precipitation from space—EURAINSAT and the future. *Adv. Glob. Change Res.*, 28, 49–58.

Yilmaz, K., R. Adler, Y. Tian, Y. Hong, and H. Pierce, 2010, Evaluation of a satellite-based global flood monitoring system. *Int. J. Remote Sens.*, 31, 3763–3782. (10.1080/01431161.2010.483489).

6 Algorithmic Design Considerations for Geospatial and/or Temporal Big Data

Terence van Zyl

CONTENTS

6.1 MOTIVATION

In order to frame the geospatial temporal big data conversation, it is important to discuss them within the context of the three Vs (velocity, variety, and volume) of big data. Each of the Vs brings its own technical requirements to the algorithmic design process, and each of these requirements needs to be considered. It is also important to acknowledge that some of the challenges facing the broader big data community have always existed within the geospatial temporal data analytics community and will always continue to do so.

Especially relevant are those big data challenges relating to data volume as presented by large quantities of either raster data, point clouds, and even vector data. Spatial data mining has long endeavored to unlock information from large databases with spatial attributes, and in these cases, algorithmic approaches have been adapted to overcome the data volume. Although the problem of big data is one that is well acknowledged and long studied, it is worth gaining a deeper insight and a more formal and rigorous treatment of the subject as is presented by the opportunity of a sudden awareness of spatial big data by the broader data community.

Spatial data can be categorized into three major forms: these being raster, vector, and areal. Historically, it has been the case that raster data presented itself as a large volume challenge. It is clear that this historical trend is changing, and none of these categories maps neatly to any of the big data's Vs. For example, a large volume of vector data is now plausible if the Internet of Things is considered, and these data could also place velocity constraints on the algorithms if near real-time processing is required. Additionally, high-variety unstructured data may arrive at high velocity or any other of the many permutations.

What is clear across all these permutations of the big data Vs is that considerable consideration needs to be given to the time and space complexity of the algorithms that are required to process these data. In addition, each of the three Vs places added constraints on the others, and increasingly, the three Vs need to be considered together. For example, unstructured data increase the time complexity of algorithms needed to process the data chunks, while, for instance, high volumes of the same unstructured data increase space complexity. To gain a true sense of the overall challenge faced by the geospatial big data community, couple these classical challenges of big data with the added time and space complexity of spatial data algorithms.

First, it is important to note that the independent identical distribution (IID) is not a reasonable assumption for either temporal or spatial data. The reason for this assumption failing is that both of these cases consider data that is autocorrelated. In fact, the first rule of geography is this fact exactly. As a result of not being able to make an IID assumption in most cases, the time complexity of spatial and temporal algorithms is higher than their traditional counterparts. For example, Spatial Auto-Regression is more complex than Linear Regression, Geographically Weighted Regression is more computationally demanding than Regression, and Co-Location Pattern Mining requiring spatial predicates is more complex than Association Rule Mining. In addition, ignoring the spatiotemporal autocorrelation in the data can lead to spurious results, for instance, the salt and pepper effect when clustering (Goodchild et al. 1992).

The solution to the big data challenge is simple to describe yet in most cases is not easily tractable. Simply put, it is important insofar as it is possible to minimize space complexity aiming for at most linear space complexity and target a time complexity that is log linear if not less. However, this is often not possible and other techniques are required (Vatsavai et al. 2012).

All is not lost and spatial data does not only present increased challenges in the big data arena but also provides additional exploitable opportunities in overcoming some of the big data challenges. For example, spatial autocorrelation allows for aggregations and filtering of data within fixed windows so as to reduce the total number of points required for consideration without excessive loss of information. It also allows the algorithm designer to consider points at sufficient distance as a single cluster thus reducing the number of computations.

6.1.1 Challenges

Many sophisticated or more complex algorithms require variable and unpredictable traversal paths, patterns, and frequencies of access to the underlying data structures. As a result of this somewhat stochastic data access patterns, cache coherency can be seriously undermined by the algorithm causing thrashing and as a result slowdowns.

Algorithms that require the same data to be reconsidered in multiple iterations or steps and that may change the same data multiple times have increased time complexity and memory requirements. When dealing with both volume and velocity challenges, these multiple iterations and alterations need to be eliminated from the algorithms either by restructuring algorithms so that they operate in a single pass or by using clever tricks to reduce the concurrent data being considered.

Another major consideration is the infrastructure that will be used to support the big data algorithms. For instance, if the algorithms are required to use MapReduce, then certain limitations on iterative algorithms come into effect. Minimizing shared resource contention is also a major consideration during the algorithmic design process for a given infrastructure. However, the selection of infrastructure platforms and toolsets made available has become vast, and almost any of the various algorithmic styles can be placed in the appropriate framework (Cary et al. 2009).

6.1.1.1 Algorithmic Time Complexity

The complexity of many spatial and temporal algorithms stems from their development for historical environments where data at each location in both space and time were sparse. The traditional approaches to overcoming this sparsity were to exploit the autocorrelation of locations in proximity to one another. This exploitation required complex algorithms with many assumptions made about the underlying structure of the data. Big data change these original considerations allowing one to consider smaller intervals or subspaces in relatively increased isolation as a result of the sheer volume of data per location at a given time.

A typical solution exploiting this ability to consider smaller subspaces is a divide and conquer algorithmic approach where algorithm designers build an individual model per subspace or do per subspace evaluations. A most basic example is large satellite time series data cubes that are often processed using per pixel algorithms.

Another approach is to reduce the complexity of the algorithm used allowing the sheer volume of the data to guide the process rather than assumptions on structure. Applying simple algorithms can be captured as a rule of thumb by stating that the bigger the data volume, the simpler the model that can be used.

Using subspaces or simpler algorithms is not a silver bullet, and often, the large amount of data results in its own challenges. For instance, vast amounts of data will as a result of random error tend to cluster in many ways, and it is often possible to find any pattern you wish. This algorithmic designer bias may result in an entity discovering the patterns they wish for rather than the ones that actually exist.

6.1.1.2 Algorithmic Space Complexity

A classical challenge presented by many algorithms is the requirement to consider the relationship between every pair of observations in order to build some quadratic space complexity matrix such as a spatial weight matrix W or some distance D or similarity S matrix. It is inevitable for a large volume of data that the size of this matrix will exceed the memory capacity of the system. This is less of a challenge in time series applications due to the 1D nature of time and the natural unique ordering on it. However, when considering spatial relationships between objects, the number of combinations is great and needs to be considered in far more dimensions. Consider as an example the case of 2D space where no natural unique ordering of objects exists (Shekhar et al. 2003).

The large data requirements are exasperated when not only measurable distance relationships such as Euclidean and Manhattan distance are considered but also semantic relationships such as connectedness or similarity need to be accounted for. In order to overcome this complex challenge of describing the weight between every data point in the data set, a N × N matrix is formed where the ith and jth element stores the weight of the relationship between those two elements. Note that this relationship is bidirectional and may have different values in different directions.

In order to overcome the large matrix W, algorithm designers use three major approaches: (first) the relationship needs to be calculated iteratively as the algorithm progresses, (second) given some algorithms that converge on the correct answer, one may consider using batches or subsets of the data to continuously refine the model parameters, or (third) the weight matrix may be mapped into some lower-dimensional space using techniques such as multidimensional scaling (MDS).

Now, consider the challenges of big data where either the data is of high volume, in which case this matrix grows quadratically as the data grows, or the case is of high velocity, in which case this matrix is continuously changing and can never be complete. In the case of unstructured data as presented by high variety, often, this matrix may not even be fully specified with the need to handle missing or incomputable values and dimensions. This is exasperated by a common need to invert the weight matrix W that at present is at least an $O(N^{2.x})$ operation. Unfortunately, the use of such a weight matrix is a common practice in many traditional geospatial statistical algorithms as the mechanism of dealing with autocorrelation, and as such, many of these geospatial statistical techniques become intractable solutions in the geospatial big data space (Vatsavai et al. 2012).

6.2 GEOSPATIAL BIG DATA ALGORITHMS: THE STATE OF THE ART

Spatial and/or temporal algorithms are required to deal with a diversity of data types including raster data cubes such as time series of satellite images; vector data as in points, lines, and polygons found in, for instance, in situ mobile data sensors; and the graphs found in, for instance, a roadmap. Given that none of these data types are unique to any of the Vs of geospatial big data, the arsenal of algorithmic approaches required is vast. Consider, for example, the following list of use cases in which spatiotemporal big data algorithms are required to be applied (Shekhar et al. 2003):

- Spatial and temporal clustering
- Spatial outlier and temporal change detection
- Spatial and temporal interpolation
- Spatial extrapolation and temporal forecasting
- Spatial and temporal classification
- Spatial co-location and temporal co-occurrence
- Spatial relational fact extraction from unstructured data

Here, we consider some of these use cases and the current state of the art.

6.2.1 VOLUME ALGORITHMS

The geospatial temporal volume problems are some of the most prevalent. Consider, for example, the size of data produced by modern satellites, weather radar, and other remote sensors. Increases in both spatial resolution and temporal repeat rate coupled with constellations of satellites mean that the volume challenge is unlikely to abate in the near future. Other high volume sources of geospatial data include unstructured data from, for instance, geo-tagged photographs and geo-located text messages.

Solutions to volume problem commonly include the following (Vatsavai et al. 2012):

- Place-based ensemble of decision trees for Land-cover classification; here, a number of simple models (ensemble) are used on subregions as a mechanism for reducing time and space complexity.
- Place-based ensemble of spatial auto-regression models also reduces computational and space complexity.
- Stochastic relaxation and iterative approaches to spatial auto-regression.
- Performance-optimizing classification of time series based on nearest neighbor density approximation.
- Markov random field with $O(N^2)$ space requirements and $O(N^3)$ computation complexity, as a result of the use of Gaussian processes, can be reduced to $O(N)$ and $O(N^2)$ through the use of clever algorithms.
- K-means, although it is an NP-hard algorithm, can, through the use of heuristic, be reduced to an $O(N\log N)$ algorithm that makes it particularly attractive.
- KD-trees have a $O(N\log N)$ time complexity that makes them particularly attractive in big data scenarios where they can be used for subsetting and filtering.

6.2.2 Velocity Algorithms

Problems of velocity come in two major classes: those dealing with data that need to be processed and decisions made as the data arrive on a per item basis (these are known as online algorithms). In contrast, streaming algorithms are also required to process data as they arrive; however, decisions are not required at each data point; instead, points may be batched and decisions differed until sufficient information has arrived. The usual constraints driving the need for streaming algorithms are limited resources in the form of memory, for instance, or limited processing time. In both instances, that is, streaming and online, the algorithms are required to model phenomenon with incomplete information.

To make the comparison more concrete, we consider a streaming algorithm that is required to test for out of bounds on water quality sensor data stream and provide an alert if a threshold is breached. In order to ensure that the threshold has indeed been breached, the algorithm may wait for future observations to confirm. The alternate is an online algorithm that tests sensor observations to evaluate their validity; if the observation's value is out of bounds, the data are discarded. Common streaming algorithms to tackle big data include the following:

- Hoeffding option tree.
- Naïve Bayes is a particularly attractive option due to its low time and space complexity.
- Stream KM++.
- D-Stream.
- Incremental linear discriminant analysis.

6.2.3 Variety Algorithms

The variety challenge comes in two forms: first, variety includes the case of unstructured and semi-structured data, and second, it considers the case of large numbers of features from many sources. In the case of semi-structured and unstructured data, the complexity comes from the need to process the unstructured data and extract usable information. In the case of large numbers of features, the features add additional time complexity, cause dimensionality challenges, and can often be correlated with one another leading to instabilities. Also, the large numbers of features may need to be merged as they are inconsistent measurements of the same phenomenon, where the observations come from different sources.

Variety is an open challenge in the big data space; many of the techniques used for tackling volumes or velocity of data can come to bare. When coupled with effective dimensionality reduction, these can help overcome many of the variety challenges. However, the spatial temporal uncertainty that comes from dealing with unstructured data is not well understood.

On the other end of this, given spatial and temporal features associated with unstructured data can often reduce some of the complexity by allowing the autocorrelation to be exploited leading to better solutions.

6.3 ANALYSIS OF CLASSICAL GEOSPATIAL AND TEMPORAL ALGORITHMS

Let us consider briefly a number of classical common geospatial and temporal algorithms used within the geospatial statistics and geospatial analytics community. Most of these algorithms have been historically developed for sparse data scenarios, such as Kriging, developed for interpolating geological properties using a limited number of core samples. For this reason, classical algorithms seldom take cognizance of the big data challenge and instead try to maximize the information that can be extracted from the relatively small sample of data provided by making appropriate assumptions (Shekhar et al. 2011).

The major constraint in many of these classical algorithms is the weight matrix W placing an $O(N^2)$ constraint on space coupled with a minimum $O(N^{2.x})$ computation constraint if this matrix is inverted with this being as bad as $O(N^3)$ in the likelihood-based scenarios.

First, some of the most common algorithms used for classification and regression are Geographically Weighted Regression and Logistic Regression. Both these algorithms have $O(N^3)$ space complexity but, through the use of various improved optimization techniques, can have their time complexity greatly reduced.

Typical spatial interpolation is done using Kriging that has an $O(N^3)$ space complexity. Gaussian elimination used in solving many of these algorithms has an $O(N^3)$ time complexity and an $O(N^2)$ space complexity; however, this can be significantly improved through some clever tricks.

Maximum likelihood estimation is a common approach in spatial and temporal modeling when considering parameterizations; however, it results in an $O(N^3)$ time complexity.

What is common across all of these algorithms is the massive space and computational complexity associated with all of them. What is clear immediately is that solving spatial temporal analytics problems will require significant manipulations of these algorithms on even reasonable size data sets (Vatsavai et al. 2012).

6.4 APPROACHES TO ALGORITHMIC ADAPTATION FOR GEOSPATIAL BIG DATA

This section presents a number of approaches to scaling algorithms for big data and hints at how these approaches might be applied to the classical algorithms presented in Section 6.3.

6.4.1 DIVIDE AND CONQUER

One important consideration when dealing with large volumes of data is that a [N × N] matrix is significantly larger than two [N/2 × N/2] matrices. This fact is exactly the effect exploited by the divide and conquer approaches. The divide and conquer approach to algorithmic design requires that the problem can be divided in smaller subproblems, each of which can be solved and from which the overall solution can be composed. In the simplest case such as those presented by problems

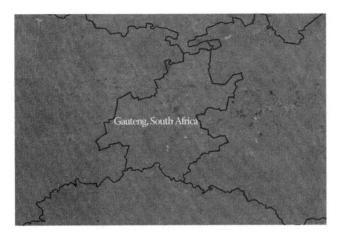

FIGURE 6.1   Use of the Mann–Kendall change detection algorithm on a per pixel basis over Gauteng, South Africa, looking at a 10-year time series of EVI values.

that have many opportunities for parallelism, it is possible to consider each part of the divided problem in complete isolation without any need for fusion of the results. An example of such a highly parallelizable technique is used by the remote sensing community when doing time series analysis on large data cubes. Instead of considering a pixel and its entire neighborhood in both space and time, the algorithms consider each spatial pixel independently and analyze the time series for just that pixel. Figure 6.1 shows the use of the Mann–Kendall change detection algorithm on a per pixel basis. Another approach is to use aerial boundaries such as geopolitical borders and even natural borders such as rivers to divide the total region into smaller regions that can then be analyzed separately.

In the case of data velocity, the same technique may be applied by dividing the data into substreams, each of which can be processed separately and has the same divide and conquer results.

6.4.2 Subsampling

Subsampling is a technique whereby a subset of the original data set is used as a representative example of the full data set. If this subsample is chosen correctly and in such a way so as not to bias the algorithms, the subsample may be sufficient to give a good enough parameterization to models. Although the model may not be as accurate as if the full data set was used, it may be the case that the loss in accuracy is minimal. In addition, as a result of the subsampling, the algorithmic time and space complexity can be reduced. Another thing to note is that additional accuracy can always be gained by considering more data. This allows the modeler to balance the risk, time, and space requirements of the algorithm so as to arrive at an answer.

Subsampling is not only relevant to the volume problem but is also effective as a technique in velocity and variety challenges. If a local stationary assumption holds on the stream of data, it is possible to model the underlying phenomenon using a sample of the data rather than the full data set. Or in the case of unstructured text, it may be possible to gain an understanding of the criteria under evaluation using some of the text rather than all of it. For instance, take a high three Vs challenge of doing sentiment analysis using geo-referenced text feeds. Here, an effective sampling of the messages as they stream past using sentiment analysis on that subsample would allow for effective hot spot detection. If, however, the challenge was to detect anomalous text messages coming from certain regions with the intention of, for instance, monitoring terrorists, the same subsampling technique would be of no or little value, since the observations being searched for are sparse and not representative of the population (Demetrescu and Finocchi 2007).

However, subsampling comes with its own caveats. Primarily, being what is the correct sample size so as to ensure that all of the underlying properties being modeled are represented. For instance, a too small sample size may lead us to believe that only a global property holds. Or if the sample is slightly skewed by the stochastic selection process, a certain local property holds in some areas that have been relatively oversampled. In the same way as it is possible to sample with bias from a population if the experiment is not carefully designed, it is possible to subsample with bias if the subsampling is not considerate of the data's underlying distribution.

6.4.3   AGGREGATION

Another common approach is to reduce data by aggregating the data. Autocorrelation allows for further opportunities for aggregation as locations close to one another are by their nature related and as such can be aggregated. In fact, for the case of geospatial data, this relationship is often a similarity relationship that creates increased opportunities for aggregation. These aggregations include averages such as mean or median and other statistical summaries such as variance and properties that describe the data's distribution.

By treating areas at a distance as a single aggregated entity, this can reduce the total number of entity to entity relationships considerably. Consider, for instance, that a logarithmic reduction of entities using aggregation could change an $O(N^2)$ algorithm to an $O(N\log N)$ algorithm. Aggregation can also have a significant effect on space complexity of an algorithm by allowing algorithms to work only with massively reduced data quantities by only using spatial or temporal aggregation in the modeling process. Aggregation has a second effect in that it often reduces the noise in the data leading to fewer spurious results.

6.4.4   FILTERING

One common approach to dealing with the challenge of big data is to filter the data. This is somewhat different from subsampling where we select a subset of the data using some stochastic criteria. Filtering allows us to select a subset of the data based on some criteria whereby we believe the subset of the data will be representative of the data required for our modeling purposes in the area of interest.

Consider, for example, a certain phenomenon that we wish to model locally given the first law of geography and with sufficient knowledge of the speed with which the phenomenon is changing; it makes sense to filter the data using some windows

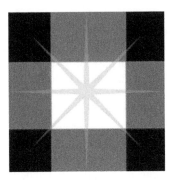

FIGURE 6.2   Center pixel with surrounding 8 pixels filtered to form a single feature.

that include only nearby points that can have an effect. An example of this is used often in the remote sensing community where a subset of pixels around a given pixel is included in the feature vector for modeling the center pixel as shown in Figure 6.2.

Big data volumes allow filtering to be done more effectively since we will always have sufficient data in our neighborhood to model with. Juxtapose this to traditional data accumulation that often left the modeler with a sparse sample of the underlying phenomenon and required algorithms that could maximize the information gained across all of these data points.

One example filtering is through the use of spatial indexes such as R-trees to gain a sense of the points that are close to one another and then use this subset of points to build the model. This technique is used extensively in big data to reduce time and space complexity and is characteristic of K-nearest neighbors algorithms.

6.4.5   ONLINE ALGORITHMS

Online algorithms are algorithms that require an immediate action per each item of input data as it arrives. For instance, consider a scenario where a message needs to be routed immediately upon arrival; here, the near real-time constraints on the data mean that as each item arrives, a route must be selected and the item processed.

The needs for online algorithms are often as a result of requirement rather than specifically due to the big data itself. Big data does, however, exasperate the situation in online scenarios. An example is high-variety data such as a stream of geo-located free text that needs to be processed as each message arrives. Here, algorithms processing the free text put significant time pressure on the processing requirements.

6.4.6   STREAMING ALGORITHMS

Streaming algorithms are similar to online algorithms; however, instead of the major requirement being the need for immediate action per item, the major consideration is limited space. In this instance, the answer may be deferred until a later stage, but given limited resources, often, some limitations on the amount of data being considered are required. An example may be, for instance, an anomaly detection algorithm such as a water quality anomaly; here, every item does not require immediate processing; however, since a continuous stream of data is being considered, memory limitation will soon become apparent, and the algorithm will need to consider only some relevant subsets of the data.

Streaming algorithms have their major challenges linked to resource constraints due to either limited time to spend on computations or limited memory to hold the data. That being said, they are a classic approach to dealing with the challenges of big data as they cover the requirements of all three Vs.

6.4.7   ITERATIVE ALGORITHMS

In the instance where memory requirements mean that an algorithm is unable to hold all the data under consideration in main memory, some of the burdens can be removed by changing the algorithm from a bulk algorithm to an iterative algorithm. In this instance, the algorithm may increase time complexity by reducing its memory footprint through the use of an iterative approach.

Usually, iterative algorithms have a second side effect in that they can often reduce cache coherence causing thrashing and reducing the performance of the algorithm as a result. However, in instances where the algorithm was intractable due to memory requirements, this might be the only solution.

Another place where iterative algorithms are a natural solution is high-velocity scenarios. Here, the data need to be considered either in an online situation where we need a result for each data item as it arrives or in streaming situations where we naturally need to consider each item as it arrives until some criteria are met or some events have transpired.

6.4.8   RELAXATION

It is often possible to consider an approximation or relaxation of a difficult problem by a *weaker* problem that is easier to solve. The relaxation can be seen as a way of reducing the total time or space complexity of an algorithm. The use of relaxation of constraints has been extensively considered in, for instance, integer programming. The same approach of relaxing the constraints from discrete entities to, for instance, real numbers can allow many algorithms to function in geospatial scenarios. Consider, for example, using a weaker notion of distance or relaxing other spatial constraints such as connectedness to just a Euclidean distance (Chandrasekaran and Jordan 2013).

6.4.9   CONVERGENT ALGORITHMS

Convergent algorithms do not arrive at a solution immediately; instead, each iteration of the algorithm or each additional data item considered improve the solution. In other words, each iteration reduces the uncertainty in the solution and as such reduces the error. Since, often, the focus of many model parameterizations or model explorations involve some optimization functions, this optimization is often the focus of convergence. Consider, for instance, classical example of replacing a closed-form solution for linear regression with a convergent algorithm such as Stochastic Gradient Decent.

The major advantage of convergent algorithms is in the case of high-velocity data where some solutions are required in near real-time or in an online fashion. Here, a poor solution in a short amount of time is more valuable than no solution at all. Consider, for instance, some anomalous weather detections; it makes little sense to have a warning after the event has passed (Kleinberg and Sandler 2003).

Given high volumes of data, it also makes sense to use convergent algorithms as they provide a mechanism to balance computational effort with risk or to balance memory usage with risk. In other words, the algorithm designer can choose the amount of error they are satisfied with relating to the amount of time spent or the amount of memory used. Given the current emergence of commodity computing in, for instance, cloud computing, this may be an important consideration.

6.4.10 Stochastic Algorithms

Stochastic algorithms use some forms of randomization as a mechanism of either finding a solution or overcoming an adversarial scenario. Stochastic algorithms take on two forms, randomized algorithms and probabilistic algorithms. Randomized algorithms such as the Quicksort are guaranteed to find the correct solution; however, they do not guarantee an upper bound on time complexity. Probabilistic algorithms, such as Monte Carlo and Las Vegas, do not place any bounds on computational complexity but can, in the case of Monte Carlo algorithms, be guaranteed to converge to a correct solution given infinite time or in the case of Las Vegas algorithms, may not even guarantee this.

Stochastic algorithms are particularly important in big data scenarios where often having some heuristics that converge on a good solution is sufficient for the given use case. Take, for example, the use of particle swarm optimization that uses a guided stochastic algorithm or a heuristic to optimize a given function but allows the algorithmic designer to decide the total computational time.

6.4.11 Batch versus Online Algorithms

The traditional approaches to spatial temporal models, especially for their use in environmental sciences, are to parameterize the model in an offline batch fashion. That is, given the area of interest and a time of interest, all data required by the algorithms in order to generate the model are made available as a single batch that can then be processed. Big data requires the consideration of streaming data that needs to be processed continuously. Streaming and online algorithms are also used in scenarios including near real-time monitoring and anomaly detection. Some examples of near real-time monitoring and anomaly detection include alerting to unfolding events such as natural disasters, diseases, crime, and unrest. These data used in near real-time monitoring and anomaly detection come from sensor data streams, simulation models, or unstructured geo-referenced text messages.

Mobile applications such as vehicle tracking also place extensive requirements on problems in the online algorithm space. These applications often have routing requirements or near real-time processing requirements. In the example of vehicle tracking, a typical anomaly detection case may involve the theft of a vehicle. Here, an immediate response is required considering the volume of data and the large number of variables, involving anything from trailer states to CO_2 emissions to detection of an anomaly, will need to be processing the variety of variables.

6.4.12 DIMENSIONALITY REDUCTION

One mechanism used to deal with the challenge of variety is through dimensionality reduction. Dimensionality reduction makes the assumption that data lies on lower-dimensional manifold embedded in the higher-dimensional space. Various techniques have been proposed, the most common being principal component analysis. However, most of these techniques have high time and space complexity. Using relaxations such as covariance tapering or subsampling on the estimations of, for instance, covariance matrices can greatly reduce the total complexity (Kaufman et al. 2008).

6.4.13 EXAMPLE

In order to make some of the earlier techniques for dealing with big data more concrete, let us consider the following examples. Although the example is somewhat contrived in a big data setting, it serves the purpose to evaluate some of the techniques' effects in terms of quality of solution with respect to execution time. The algorithm being used is a variable kernel width density estimation algorithm to produce a density estimate for active fires. Figure 6.3 shows a typical output for the algorithm.

What this means is that a unique bandwidth is used for each point in the kernel density estimate as opposed to the traditional use of a single bandwidth. The algorithm used to calculate the bandwidths is the Maximum Likelihood Leave-One-Out Kernel Density Estimator. To find the kernel width for all N elements using this algorithm is of order $O(N^2 + D^3)$ time complexity with an order $O(N^2 + N.D^2)$ space complexity.

First, in order to make the algorithm tractable, the space complexity is reduced by using an iterative version of the algorithm that calculates the distance between points as they are needed that results in a reduction to an order $O(N.D^2)$ space complexity.

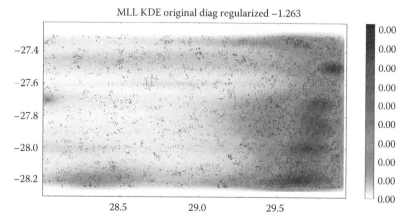

FIGURE 6.3 Kernel density estimation for active fire incidents per year over 10 years (2002–2012). Dark gray indicated a higher number of incidents per year.

TABLE 6.1
Adaptations to the Kernel Density Estimation
Algorithm Are Shown along with the Speed
Up Obtained and the Related Log Likelihood
Score as a Result of the Adaptation

Adaptation	Time (s)	Log Likelihood
Iterative	38335.44 (10.64 h)	−1.196
Relax constraints	139.69	−1.225
Relaxed convergence	70.67	−1.263
Subsampling	45.21	−1.265

In order to reduce time and space complexity, some constraints are relaxed and the requirement that the bandwidths for the kernel density estimator be represented by a full covariance matrix is reduced to that of a diagonal matrix, which has the effect of removing a costly matrix inversion. As a result, the time complexity is reduced to order $O(N^2)$ and space complexity is of order $O(N.D)$. Now, since an iterative algorithm is used, it is possible to relax the convergence criteria and allow the volume of data to guide the process requiring that only one iteration of the stochastic gradient decent optimization step be done for each data point.

Next, consider the case where subsampling is used to select a subset of points for the calculation of the bandwidth. If some fractions $f = 0.3$ of the samples are subsampled, the time for the bandwidth calculation can be reduced; note this does not change the time complexity. In order to overcome any adversarial conditions, the subsample is selected stochastically for each bandwidth to be calculated, allowing the volume of data to guide the convergence of bandwidths in a given region to the correct value.

Table 6.1 shows the results for various runs of the algorithm using the techniques described here in order to make the algorithm more tractable in a big data setting. In this example, what is noticeable is that the use of subsampling has almost no impact on the accuracy (likelihood score) with a marked decrease in execution time. Notice the massive time gained by relaxing the constraints on the algorithm and allowing it to work in a diagonal setting. Although it is true that some of the techniques result in some loss of accuracy, given that a solution may be acquired in 45 s rather than the 10 h, this loss in accuracy may be acceptable.

6.5 OPEN CHALLENGES

Some of the major open challenges that still need exploration when considering some of the strategies considered here to overcoming the big data challenge include the following:

How do ensembles of simpler models compare with current spatial models, and can we fully describe the loss function when moving from one to the other as a function of the data? In short, when do bigger data allow the use of ensembles of simpler models?

When can we use filtering effectively under which scenarios and which mathematical and statistical assumptions? How much filtering is allowed before the solution completely degrades and becomes nonrepresentative of the actual solution? Can we fully describe the relationship between the amount of filtering being applied, the amount of data, and the quality of the solution? What types of filtering are allowed under which types of data?

It is well understood that aggregation is not only an effective form of data reduction but also may be applied to any data leading to very wrong results. For instance, making an assumption about the mode of the data and applying a given aggregation can lead to complete wrong models. In all instances of big data, often, a clear mechanism for data exploration and statistical property discovery is not clear, and this is reinforced in the spatial and temporal domains. Consider, for example, high-velocity data where the data may be non-stationary, and by its very nature, these aggregations will be wrong or can only be correct on some local temporal scales.

Divide and conquer as an approach to big data often neglects spatial autocorrelation as is the case in a per pixel time series analysis. When is it safe to do so and how much information is being lost as a result of this approach?

One challenge especially in the variety space that has not been considered is that of privacy. Geo-location of free text or images can quickly infringe on people's privacy, and this needs to be considered even if the data has been anonymized.

Variety as in unstructured data mining for geo-co-location is an open challenge highly dependent on our capability to tackle the challenges of AI and, in so doing, understand the meaning of free text or on our capability to extract meaning from, for instance, images or videos.

Another challenge relating to streaming and online algorithms is if the solutions to various models can be computed precisely using online algorithms, and if they cannot, what is the error in the approximations?

6.6   SUMMARY

An evaluation of the techniques that can be used by algorithm designers to make the challenge of geospatial and temporal big data more tractable has been presented. The techniques presented include divide and conquer, relaxed constraints, filtering, subsampling, dimensionality reduction, online algorithms, iterative algorithms, aggregation, streaming algorithms, and stochastic algorithms. When these techniques are used together or in conjunction with big data tacking cognizance of the opportunities presented by the spatial characteristic of the data, massive gains can be made in reducing time and space complexity.

It is safe to say that strides are being made toward solving the challenges of geospatial and temporal big data; however, many open challenges exist and no silver bullet will ever exist. What is clear at this stage is that insufficient formalism and understanding of the relationship between big data, time and space complexity, and the risk associated with the resultant relaxed or approximated solutions that arise from the algorithms are used to effectively deal with the big data challenge.

REFERENCES

Cary, A. et al., 2009. Experiences on processing spatial data with mapreduce. *Scientific and Statistical Database Management*. Springer, Berlin, Germany.

Chandrasekaran, V. and M. I. Jordan, 2013. Computational and statistical tradeoffs via convex relaxation. In *Proceedings of the National Academy of Sciences*, 110(13): E1181–E1190.

Demetrescu, C. and I. Finocchi, 2007. Algorithms for data streams. *Handbook of Applied Algorithms: Solving Scientific, Engineering, and Practical Problems*, Nayak, A. and I. Stojmenovic (eds.). John Wiley, Hoboken, NJ, p. 241.

Goodchild, M., R. Haining, and S. Wise, 1992. Integrating GIS and spatial data analysis: Problems and possibilities. *International Journal of Geographical Information Systems* 6(5): 407–423.

Kaufman, C. G., M. J. Schervish, and D. W. Nychka, 2008. Covariance tapering for likelihood-based estimation in large spatial data sets. *Journal of the American Statistical Association* 103(484): 1545–1555.

Kleinberg, J. and M. Sandler, 2003. Convergent algorithms for collaborative filtering. In *Proceedings of the 4th ACM Conference on Electronic Commerce*. ACM, New York.

Shekhar, S., M. R. Evans, J. M. Kang et al., 2011. Identifying patterns in spatial information: A survey of methods. *Wiley Interdisciplinary Reviews: Data Mining and Knowledge Discovery* 1(3): 193–214.

Shekhar, S., P. Zhang, Y. Huang, and R. R. Vatsavai, 2003. Trends in spatial data mining. *Data Mining: Next Generation Challenges and Future Directions*. AAAI/MIT Press, London, U.K., pp. 357–380.

Vatsavai, R., A. Ganguly, V. Chandola et al., 2012. Spatiotemporal data mining in the era of big spatial data: Algorithms and applications. In *Proceedings of the 1st ACM SIGSPATIAL International Workshop on Analytics for Big Geospatial Data*. ACM, Redondo Beach, CA.

7 Machine Learning on Geospatial Big Data

Terence van Zyl

CONTENTS

7.1 MOTIVATION

When trying to understand the difference between machine learning and statistics, it is important to note that it is not so much the set of techniques and theory that are used but more importantly the intended use of the results. In fact, many of the underpinnings of machine learning are statistical in nature. When considering statistics, the main intent of statistics is in gaining an understanding of the underlying system, in this case geospatial system, through an analysis of observations or data about the system. Here, the geostatistician or environmental modeller is interested in cause and effect in the underlying system and gaining a deeper understanding of system itself. As a result of the need for environmental modellers and geostatisticians to gain an understanding of the underlying system, it is important that the eventual statistical model be interpretable, that is, not a black box. In fact, one reason for the limited use of machine learning algorithms has historically been exactly the lack of interpretability.

Machine learning, on the other hand, is more focused on learning from observations of a system so as to be able to automate functionality. Here, the intention is not one of understanding but more one of engineering. For instance, in machine learning, a model may be trained so as to do automated classification of new unlabelled observations, to forecast future observations of some system or automatically spot anomalous events (Vatsavai et al. 2012).

Geospatial big data present two opportunities for the increased use of machine learning in the geospatial analytics domain. First, geospatial big data have created a shift toward considering large amounts of data as a resource that can be used to add value to an organization. Second, by virtue of the three Vs, volume, velocity, and variety, of big data, there is a shift away from complex models that require extensive computational and memory resources to techniques that instead can produce results in a more computationally efficient manner. Both of these opportunities provide a space in which black box solutions that produce *usable* results are more valuable than a strict need for interpretability and transparency.

Machine learning has historically been designed and tested on small and medium data instances with a moderate to large number of attributes or features. The main assumption of machine learning is that the data are independent and have identical distributions. Machine learning is not, however, a panacea and the three Vs of big data result in a number of challenges to machine learning. The machine learning challenges include the following:

- Big data volumes resulting in the need to deal with numbers of training examples beyond current in-memory processing capability
- Big data variety creating feature dimensionality beyond computing capability
- Big data variety increasing learning complexity so as to not be doable in a meaningful time frame
- Big data velocity placing requirements on the time efficiency of both learning and classification which are not attainable

Geospatial big data being inherently autocorrelated, both spatially and temporally, add to the previous challenges. For instance, current solutions looking to exploit the aforementioned autocorrelation use either windowing approaches or more sophisticated machine learning algorithms such as recurrent neural networks. This increases the challenges to machine learning created by big data in the following ways:

- Use of windowing increases dimensionality.
- Including spatial relationships increases dimensionality.
- Sophisticated geospatial machine learning algorithms increase learning complexity.

The current approach to big data in the machine learning community is to use a large number of very simple learners and then to allow the sheer volume of data to guide the learning. Although, this approach works very effectively on geospatial data, it is not without faults as it ignores the spatial and temporal autocorrelation to some extent. Additionally, large amounts of data mean large amounts of noise, and by its very nature, this tends to cluster, generating spurious patterns and complicating the machine learning process. These challenges are all amplified in the geospatial context where errors can be autocorrelated (Gilardi et al. 2003; Vatsavai et al. 2012).

7.1.1 Supervised, Unsupervised, and Feature Learning

There are three major areas of machine learning that apply to the task of geospatial data analytics under consideration in this chapter: (1) supervised learning, (2) unsupervised learning, and (3) feature learning.

7.1.1.1 Supervised Learning

Supervised learning is primarily concerned with classification, interpolation, and prediction. That is, in the case of supervised learning, the learner is given a set of training examples or input features that represent an observation of the values of some phenomena; in the case of geospatial big data, one or more of those features may be spatial in nature. The supervised learner is also given a label for each training example. For instance, the features of the training example could be water quality values, and the labels could be drinkable and not drinkable. It is the function of the learner to learn the mapping or classification from the training examples to the output labels. Once this mapping has been learned, the learner can then be used to label unlabelled examples.

In the case where the set of labels being learned are a continuous set of values, the learning process is termed regression rather than classification (Mitchell 1997). Since the learned label could be a future value, it is possible to use supervised machine learning for prediction. In the case where we wish to obtain a label for an unlabelled spatial location, supervised learning can be used to do interpolation. Also, since many machine learning techniques are sophisticated and capable of dealing with nonlinearities, they are especially useful when dealing with the big data challenges of variety (Kotsiantis et al. 2007).

7.1.1.2   Unsupervised Learning

Unsupervised learning, on the other hand, is primarily concerned with clustering or grouping of examples and can be used effectively for anomaly detection and is also an effective preprocessing step as in the case of dimensionality reduction. The main purpose of unsupervised learning is to find some structures to unlabelled training examples. Once a structure is found, it can be exploited in a number of ways. First, it can be used to aggregate the data in some way arriving at a subset of *representative* or generalized examples. Second, any new examples that do not fit the learned structure can be marked as anomalous and dealt with appropriately. Third, the structure may point to a mechanism of representing the data in a more compact way such as the case of dimensionality reduction or parameterization of some density estimation. In the case of geospatial big data, adding spatial features to training examples is natural as an additional set of dimensions along which clustering can take and its use is validated by the first law of geography.* It should be noted that many of the functions of unsupervised learning such as aggregations, representative examples, dimensionality reduction, and density estimation are powerful tools in addressing many of volume challenges of big data (Ng et al. 1994).

7.1.1.3   Feature Learning

Feature learning is usually a preprocessing step and is focused on the extraction of features for learning purpose from the data. For instance, given a set of training examples each with a set of explicit features, do there exist other implicit features within the training examples that can be learned and extracted to present either the supervised learning or unsupervised learning techniques? The purpose of learning and extracting these features is to present them to further machine learning steps including further feature learning steps that may be able to use these implicit extracted features to improve their learning. Geospatial feature learning is a relatively new field as can be seen by the popularity in Table 7.3 that is, however, likely to hold much promise in the future (Coates et al. 2010).

For any of these types of learning, if the learning is done in such a way that the learning increases as each example is presented rather than over batches of examples, the learning is said to be online. Online supervised, unsupervised, and feature learning are particularly valuable when dealing with the big data challenges of velocity.

7.1.2   Big Data Challenges

The geometric interpretation of data used within many machine learning techniques, especially the use of Euclidean distance as a measure of similarity, lends itself well to geospatial vector data, and often the autocorrelation of locations close to one another as encapsulated in the first law of geography is naturally captured in many machine learning techniques and requires no modifications. Spatial data are less concerned

---

* Everything is related to everything else, but near things are more related than distant things.

with the absolute location of observations and more concerned with the relative distance between observations. In many instances, this relative spatial relationship is all that is required by many machine learning techniques to function affectively. However, this may not lend itself well to certain machine learning techniques that often require normalized transformations of the data that can result in distortions of the distance between observations and misrepresent the autocorrelation relationships (Vatsavai et al. 2012).

Another challenge is that often the spatial relationship between geospatial observations is captured by a $N \times N$ spatial weight matrix W shown in Table 7.1, which holds either the spatial distance, connectedness, or similarity between pairs of observations. For instance, this may be the shortest distance along roads between pairs of geographic locations. In contrast, a large number of machine learning techniques work in a feature space where each observation consists of a feature vector in R^D dimensions, where each element of the vector is an attribute of the observation. An example of such a feature vector is shown in Table 7.2. Here, the distance between observations is captured by a metric on the feature space such as Euclidean distance in Euclidean Space.

TABLE 7.1

$N \times N$ Weight Matrix W Where the Value of Item W_{ij} Contains the Weight of the Relationship between the ith and jth Observations

W_{00}	W_{01}	W_{02}	W_{0N}
W_{10}	W_{11}	W_{12}	W_{1N}
W_{20}	W_{21}	W_{22}	W_{2N}
—	—	—	—
W_{N0}	W_{N1}	W_{N2}	W_{NN}

Note: Weight may not be symmetrical.

TABLE 7.2
7D Feature Vector

Age	Sex	Temperature	Rain	Location X	Location Y	Affliction Flu
10	M	23	Y	23.134	23.135	True
11	M	22	N	12.145	12.146	False
—	—	—	—	—	—	—

Note: The last dimension is the category feature that we may wish to learn based on the other features.

Some machine learning techniques especially those in the unsupervised learning group can work directly with the spatial weight matrix W. Here, what is needed is an inclusion of the nonspatial features of the observation into W and in so doing to ensure that the appropriate weighting of each feature relative to each other is maintained. There is no hard and fast rule of how this should be done, and often the solution is one of art rather than science and comes with trial, error, and experience. It is at this point that it must be noted that using a $N \times N$ data structure in the big data challenge relating to volume, using a weight matrix, is not a viable solution as the space complexity is of order $O(N^2)$ (Vatsavai et al. 2012).

On the other hand, in many tasks involving, for instance, a distance or similarity in latitude/longitude, it is often possible to calculate the pairwise distance between observations at run time and reduce the space complexity of the algorithm in so doing. Even other connectedness relationships, such as intersection or adjacency, can be calculated at run time. Being able to calculate connectedness relationship at run times means that the geometries are required to be stored as an additional dimension of the feature vector. Having these geometries as part of the feature vector, however, requires special modifications to machine learning algorithms using these values when doing inter-feature similarity or distance calculations. An alternate solution is to instead use the weight matrix W with multidimensional scaling (MDS) to transform the weights from $W \rightarrow R^N$ and add them to feature vector resulting in a R^{N+D}-dimensional feature vector.

7.1.3 Three Vs

7.1.3.1 Volume

Consider first the challenge of large volumes of geospatial temporal data, specifically remote sensing satellite data. Machine learning has historically been an effective tool in tackling the challenges of classification and clustering in large volumes of image data including geospatial raster data. Machine learning has also been used effectively in many time series econometric forecasts and other geospatial vector data tasks of classification and prediction on large volumes of data. Given its history, it is expected that machine learning will continue to be an effective tool in tackling the challenges of geospatial big data.

7.1.3.2 Velocity

Data velocity places certain user requirements on machine learning algorithms that depend on the eventual usage of the system. If the purpose is to train a classifier to label incoming data, then the machine learning algorithm can be trained offline using a subset of the data and the trained model used on new examples as they arrive. In this example, the time and space complexity of the machine learning algorithm's training is not significant but rather the time complexity of its usage. On the other hand, if the machine learning is to be used for something like anomaly detection where continuous updating of learning in an online fashion is required, then both the training and usage complexity need to be computationally economical. More formally, these two cases can be stated in terms of the stationary of the underlying process driving the observations of the phenomenon. If the process is stationary, then offline learning is possible and often only a subsample is required for this learning. If the process is not stationary, then the learning needs to be continuous and online learning is required.

7.1.3.3 Variety

Consider unstructured data of the form presented by, for instance, text. It is possible to use supervised learning to learn to geocode text documents based on, for instance, their content. Machine learning is an effective tool for working with unstructured data and some techniques such as decision trees that can be used effectively on data of different types, ordinal, nominal, etc. Machine learning has also often been applied to exactly the challenge of high dimensional data as presented by a large variety of data from many sources. Since the body of knowledge here is extensive and given the extensive use of machine learning in big data challenges related to variety, it is expected and has been shown that machine learning is an effective tool for dealing with geospatial big data variety.

7.2 GEOSPATIAL *BIG DATA* FEATURE LEARNING

One of the emerging machine learning paradigms is that of feature learning or deep learning. Consider that historically what was needed to make machine learning work in a domain was an expert who understood the domain. The expert would select the appropriate features and understand the types of transformations required to construct useful features and the appropriate machine learning techniques that could work on the selected features. This need for domain expertise was especially prevalent in the geospatial community where such individuals were even called remote sensing experts, for instance. Feature learning intends to take machine learning to the next level by allowing the processes of feature construction to be automated, in effect to be learned. This is an important step in realizing the full capability of machine learning.

Big data present two opportunities for feature learning since, first, one of the prerequisites of feature learning is large amounts of data and, second, the high volume (instances and dimensions) and high variety of big data make the prospect of a single domain expert capable of constructing features from the data intractable. Although, the techniques of feature learning have been applied extensively in other domains especially on large time series of rasterized big data, in this case videos of cats, they have not been applied as extensively to large amounts of geospatial data. Given the massive similarity between the requirements for feature extraction from videos and some forms of geospatial data, there is a good indication that feature learning techniques will be highly effective (Coates et al. 2010).

What is now emerging and presents an opportunity for research is the extent to which learned features can be rotationally invariant since there is good indication that they can be translationally invariant. This is less of an issue for spatial vector data than it is for raster data. What is needed is a solid theory of rotational invariance. What modifications should or need to happen in order to make this a capability of feature learning?

TABLE 7.3

Time Complexities and Space Complexities along with Popularity of Various Machine Learning Algorithms Used in the Geospatial Community

Technique Name	Training Complexity	Space Complexity	Usage Complexity	Geospatial Popularity[a]
Supervised learning algorithms (classification)				
Artificial neural networks	$O(D.N.K.I)$	$O(N.D)$	$O(D.K)$	107 k
Naive Bayes	$O(N.D)$	$O(N.D)$	$O(D)$	6.5 k
Decision trees (C4.5)	$O(D.N \log N)$	$O(N.D)$	$O(D \log N)$	39.5 k
Logistic regression	$O(N.D^2)$	$O(N.D)$	$O(D)$	236 k
Support vector machines	$O(D.N^3)$	$O(N.D^2)$	$O(D)$	53.5 k
Hidden Markov models (EM)	$O(K.N.I)$	$O(N.D)$	$O(D)$	31 k
K-nearest neighbor	$O(1)$	$O(N.D)$	$O(D.N \log N)$	21 k
Bayesian networks (EM)	$O(K.N.I)$	$O(N.D)$	$O(D)$	24 k
Linear regression	$O(N.D^2)$	$O(N.D)$	$O(D)$	1502 k
Gaussian process learning	$O(D.N^2)$	$O(N.D)$	$O(D)$	5.5 k
Meta (ensemble) learning algorithms				
Boosting (AdaBoost)	$O(N.R.I.F(X))$	$O(F(X))$	$O(G(X))$	8.5 k
Bagging	$O(N/R.F(Q))$	$O(F(X))$	$O(G(X))$	18 k
Random forest	$O(K.F(Q))$	$O(F(X))$	$O(G(X))$	8.5 k
Unsupervised learning algorithms (clustering)				
Self-organizing map	$O(N.D + K^2)$	$O(N.D + K^2)$	$O(K.D)$	24 k
K-means	$O(N.D.K \log N)$	$O(N.D)$	$O(K \log D)$	47 k
Mixture models (EM)	$O(K.N.D.I)$	$O(N.D)$	$O(K.D)$	35 k
Hierarchical clustering	$O(N^2.D^2)$	$O(N.D)$	$O(K \log D)$	42 k
Unsupervised learning algorithms (dimensionality reduction)				
PCA	$O(N.D^2)$	$O(N.D^2)$	$O(D)$	1.5 k
Kernel PCA	$O(N.D^2)$	$O(N.D^2)$	$O(D)$	1.5 k
Isomap	$O(N^2(D + \log N))$	$O(N.D^2)$	$O(D)$	1.5 k
Linear discriminant analysis	$O(N.D^2)$	$O(N.D^2)$	$O(D)$	13.5 k

TABLE 7.3 (continued)

Time Complexities and Space Complexities along with Popularity of Various Machine Learning Algorithms Used in the Geospatial Community

Technique Name	Training Complexity	Space Complexity	Usage Complexity	Geospatial Popularity[a]
Unsupervised learning algorithms (density estimation)				
Kernel density estimation	$O(N)$	$O(N.D)$	$O(N.D)$	9 k
Adaptive KDE	$O(N.D^2)$	$O(N.D)$	$O(N.D)$	0.5 k
Feature learning algorithms				
Autoencoder	$O(D.N.K)$	$O(N.D)$	$O(D)$	0.5 k
RBM	$O(N.D.K)$	$O(N.D)$	$O(D)$	0.5 k

Note: EM, expectation maximisation.

[a] Use the query string: (technique name) environmental or *remote sensing* or satellite or geospatial in Google Scholar to assess the relative popularity of these techniques up to 2013.

7.2.1 Approaches to Big Data Feature Learning

The most prominent feature learning techniques are those of restricted Boltzmann machines and autoencoders. Both techniques are artificial neural network techniques that use an unsupervised learning paradigm to perform feature learning. The input layers are presented with training examples, and an appropriate weight update rule is applied. The eventual purpose is to arrive at a set of new features that have been learned and that correspond to some intrinsic properties of the original data. The computational time and space complexity of both techniques is linear (see Table 7.3) making them ideal for big data applications. In the future, geospatial feature learning is likely to become an invaluable tool in the big data machine learning toolbox.

7.3 REDUCING DIMENSIONALITY OF GEOSPATIAL *BIG DATA*, MAKING MACHINE LEARNING TRACTABLE

7.3.1 Feature Construction

Feature construction is concerned with the creation of additional features to be included in the feature vector of the training examples to allow machine learning to be more effective. Better results can be achieved by presenting machine learning techniques with features that capture the properties underlying the phenomenon or features that overcome the limitations of the learning technique. For instance, a linear machine learning technique might be presented with nonlinear features by transforming the features in such a way so as to make them linear. In the geospatial context, one of the most prominent feature construction techniques is through the use of windowing.

7.3.1.1 Windowing in Raster Data

Raster data in a geospatial context are a tessellation of the plane P often 2D into a number of connected cells. A typical example of raster data is the pixels of an image such as satellite data. When considering some location $P_{i,j}$ of the raster, windowing takes into account neighboring cells shown in Figure 7.1. The function determining which neighboring cells to consider and how they should be weighted and transformed is called a window function. In the simplest case, a four-adjacency window would consider the four pixels: left, right, above, and below a pixel as shown in Figure 7.1. More complex windowing functions can be used to consider further pixels or even transformations of combinations of pixels and other properties such as how edges of the raster or missing values should be dealt with.

Windowing of raster data has two main effects: First, it allows the autocorrelation of neighboring spatial and temporal data in, for instance, a raster cube (time series

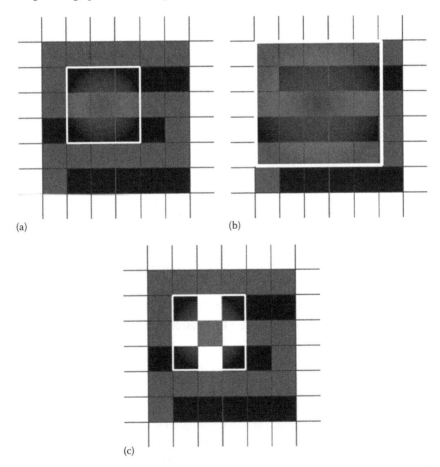

(a)   (b)

(c)

FIGURE 7.1   (a) A typical windowing function centered on a pixel in the middle of the white box. The window could apply any function to the surrounding pixels. (b) A larger windowing function centered on the same pixel. (c) A very simple windowing function centered on the same pixel and considering only the left, right, above, and below pixels.

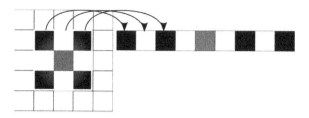

FIGURE 7.2   A diagram showing how the pixels in the window may be transformed into the features of a feature vector using a simple 8-neighborhood windowing function.

or even latitude, longitude, altitude, and time of raster data) to be captured as part of the feature vectors of the machine learning training examples. Second, it allows the $N \times M$ raster data to be transformed into a $D \times 1$ feature vectors, as shown in Figure 7.2, that can be used as the training examples to those machine learning techniques that require them in that format.

The windowing transformation, however, is not without its fair share of issues. First, the resultant feature vector is not rotationally, translationally, or scale-invariant. In fact, one of the major exercises in machine learning on raster data is constructing at a set of rotationally, translationally, or scale-invariant features by making appropriate transformation of the neighborhood matrix resulting from windowing. For example, local binary patterns are an example of one such transformation. Second, determining the optimal size or parameters of the window function so that the extracted subset of data captures the most relevant autocorrelated features needed in machine learning is also a challenge. Then, finally, the third challenge is that windowing results in any global structure in the raster being lost and it is not clear how this can be maintained while considering local structure concurrently.

7.3.1.2   Windowing in Time Series Geographic Data

Now, consider a time series of in situ observations of some phenomenon such as water temperature. Further, consider that the sensor making these observations is not stationary but instead is mobile. Here is a scenario in which there is both spatial and temporal autocorrelation of observations. In order to capture this autocorrelation in the feature vectors, a window function is used that determines which neighboring observations features should be included as additional features of the feature vector. A typical windowing function that may be used is a moving average, which weights closer in time observations more than further ones as shown in Figure 7.3. Similar

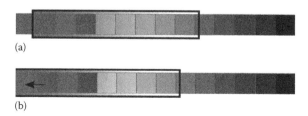

(a)

(b)

FIGURE 7.3   (a) A window selection from a time series and (b) a left shifted version of the same window function.

exponential or linear weightings can be applied to spatial distance or similarity. For instance, one might consider only the k-nearest neighbors when constructing the feature vector. Also, for instance, spatial autocorrelation may be captured by a windowing function that produces just a single value to be added to the feature vector such as the local gradient.

In general, a number of immediate challenges arise with the use of windowing; what is the appropriate window size to capture the autocorrelation, and is windowing even meaningful, for instance, the use of windowing in unsupervised learning is not valuable as has been shown theoretically? Also, although windowing does ignore global structure, local models often make sense within the context of geospatial machine learning given spatial and temporal autocorrelation.

One of the major challenges with windowing is a lack of unified theory of windowing. What size and parameterization of the window function should be used with relationship to variability of the underlying data, global and local properties of the underlying structure of the data, or the specific phenomenon being learned?

7.3.1.3   Big Data Feature Construction

The major challenge of big data with respect to geospatial feature construction using windowing is that it increases the dimensionality of the feature vector. For instance, taking into account some 20×20 window of neighborhood pixels can increase the feature vectors dimensionality by 400. Considering that satellite data resolution is increasing and that the relationship here is quadratic in space, or in the case of a time series of satellite data cubic, the space complexity can grow very rapidly using this technique. When considering the big data challenge, this increase in the dimensionality results in an increase in both volume and complexity.

7.3.2   Dimensionality Reduction

Consider again the three Vs as challenges of big data. Specifically with the challenge of variety, one of the immediate consequences of an increase in the variety of data is an increase in the dimensionality of data. This increase in dimensionality is exasperated by, for instance, less structured data that require specialized encoding to enable machine learning on the data or, for instance, by the special encoding required by windowing functions to capture the autocorrelation in time series and spatial data.

The increase in dimensionality resulting from variety has an impact on both the challenges of volume and velocity. With regard to the challenge of volume, the impact is seen as an increase in total quantity of data. In the velocity space, the impact comes as a result of increasing the complexity of the algorithms needed to deal with the added dimensions and in so doing reducing their performance. Also, in the space of machine learning, the increase in dimensionality results in a polynomial increase in the hypothesis search space; this is referred to as the curse of dimensionality. For this reason, dimensionality reduction, a technique in the unsupervised learning toolbox, is an important step in many machine learning endeavors. The main focus of dimensionality reduction is selecting the subset of features that characterize the structure of the data most appropriately. Dimensionality reduction has two major foci: those being feature selection and those being feature extraction.

7.3.2.1 Feature Selection

The focus of feature selection is to filter out a subset of features for the purpose of machine learning. The main assumption is that a subset of the features either have no relevance on the machine learning task or are redundant and already captured in the other features. Techniques such as down sampling can be effective as mechanism for reducing the dimensionality of data by selecting a subset of features that maximizes the removal of redundant information. On the other hand, feature extraction tries to transform the data from some higher dimensional space into a lower dimensional representation while minimizing any loss of information.

7.3.2.2 Feature Extraction

The primary assumption of feature extraction is that often the data lie on some manifold embedded in a higher dimensional space. The manifold may, for instance, be assumed to be linear or nonlinear, and appropriate techniques are available for each. In fact, the techniques of dimensionality reduction are well known to the geospatial data community where mapping from a higher 3D space onto a lower 2D space is common practice, which can increase performance and simplify calculations. In fact, the treatment of these reductions in the geospatial community is far more sophisticated with extensive acknowledgment of the distortions on the underlying features as a result of these re-projects of the data.

7.4 ALGORITHMIC APPROACHES TO MACHINE LEARNING OF GEOSPATIAL *BIG DATA*

Considering Table 7.3 of typical machine learning algorithms used by the geospatial temporal research community, the time and space complexities used here are typical values and do not necessarily reflect some of the more arcane or state-of-the-art variations. In Table 7.3, the name of the technique, its typical training complexity, space complexity, and usage complexity are given for demonstrative purposes (Gilardi et al. 2003; Kon et al. 2006; Lim et al. 2000). The geospatial popularity is a meta-analysis of the frequency of usage historically and does not indicate which techniques are trending upward in popularity.

What is immediately clear if linear regression is ignored that logistic regression and principal component analysis (PCA) are by far the most popular machine techniques demanding more than 50% of the total usage. Beyond the demanding usage share of those two techniques, artificial neural networks and support vector machines are the most popular supervised learning techniques, and hierarchical clustering, K-means, and mixture models demand equal shares of the unsupervised learning techniques. Here, we consider N patterns with D dimensions in a dataset X, I iterations, and K outputs; F($\mathbf{X}$) and G($\mathbf{X}$) are the training and usage complexity of an algorithm given a dataset $\mathbf{X}$; and $\mathbf{Q}$ is a subset taken from the dataset $\mathbf{X}$ with R elements. Here, usage complexity considers the case for a single input vector.

7.4.1 SPACE COMPLEXITY

In order to overcome the space complexity requirements that arise when dealing with big data and most machine learning techniques that work on batches of training examples, what is required is modifications to these techniques that instead are able to learn by evaluating individual examples incrementally. Machine learning that is done in this incremental manner is known as online machine learning.

7.4.1.1 Online Learning

Online learning allows for faster learning of an approximately correct solution by evaluating each learning example once, that is, a space requirement of effectively $O(1)$ since at any given time, only a single sample is required to be loaded into memory. Since for almost all machine learning techniques, there exist either an online version of the algorithm or it is possible to replace the optimization technique used within the algorithm with one that is iterative, it is possible to consider them as $O(1)$ space complexity algorithms although some may be quadratic $O(D^2)$ in terms of there dimensionality. This quadratic space complexity in terms of dimensions is an example of, for instance, why support vector machines have limitations in their application in the geospatial big data space.

Artificial neural networks are generally easier to train in an incremental manner than, for instance, decision trees. For this reason, in scenarios where online machine learning is required, artificial neural networks are often a natural choice despite the fact that they perform at best as well as decision trees.

7.4.2 TIME COMPLEXITY

Time complexity is an important consideration when dealing with big data and machine learning as it directly impacts on all the Vs. This immediately points to those algorithms most likely to be effective in a geospatial big data context. In fact, most machine learning algorithms are of order $O(N)$ in training examples, and often the limiting factor is in the dimensionality (Vatsavai et al. 2012).

7.4.2.1 Online Learning

Online learning often allows the time complexity of both the training and usage of machine learning algorithms to be reduced. This means that neither the velocity challenge of, for insistence, near real-time anomaly detection nor the volume challenge is insurmountable. In the velocity challenge, online learning allows for continuous learning as new training examples become available. On the other hand, when considering the challenge of volume, it is possible in many instances to do away with the repetitive revisiting of the data required by some algorithms and just allow the sheer volume of data to guide the learning process in a single pass. For instance, using stochastic gradient decent in an artificial neural network context can allow an artificial neural network to be trained in a single pass of the data, that is $O(N)$, using $O(1)$ space complexity (). From a geospatial big data perspective, online learning will work in much the same way as it does for other big data (Bottou 2010).

7.4.2.2 Ensemble Learning

Ensemble learning, shown in Table 7.3, overcomes the challenges of big data in another way. Here, the idea is to use a large number of simple learners on subsets of the data; the simple learners are then combined to form a more powerful learner. The ensemble learner is a meta-learning algorithm that describes how the data should be subsetted and how the simple learners should be combined. The data can be subsetted in two ways, either by splitting up the set of training instance into multiple sets or by splitting each training instance into multiple instances (Wang et al. 2009). Ensemble learning is very effective in a geospatial big data context where it has been used extensively for clustering. Also, given the geospatial context of the data, one technique that can be effective in ensemble methods is the natural or guided subsetting of training instances by location and in so doing exploiting the autocorrelation or redundancy in data close to one another. Random forests present one of the state-of-the-art ensemble learning techniques with many successes in geospatial big data space.

7.5 CONCLUSIONS

It is clear that the techniques of machine learning provide ample opportunity for dealing with the challenges of geospatial big data. Despite the fact that many of the newer techniques related to feature learning have not been fully explored, it is clear that techniques from the meta-learning stable such as random forests and other ensemble methods have gained extensive traction and have been used extensively. Geospatial big data machine learning is not a conquered challenge, and much of the intricacies and additional complexities of geospatial data are not completely understood or exploited within the framework of machine learning. This, however, can only provide better results than have already been achieved when they are more fully understood and integrated.

REFERENCES

Bottou, L. 2010. Large-scale machine learning with stochastic gradient descent. *Proceedings of COMPSTAT'2010*. Princeton, NJ: Physica-Verlag HD. pp. 177–186.

Coates, A., H. Lee, and A. Y. Ng. 2010. An analysis of single-layer networks in unsupervised feature learning. *Ann Arbor*, 1001: 48109.

Gilardi, N. et al. 2003. Comparison of four machine learning algorithms for spatial data analysis. *Mapping Radioactivity in the Environment-Spatial Interpolation Comparison*, Office for Official Publications of the European Communities, Luxembourg, Vol. 97, pp. 222–237.

Kon, M. A. et al. 2006. Complexity of predictive neural networks. *Unifying Themes in Complex Systems*. Berlin, Germany: Springer Berlin Heidelberg. pp. 181–191.

Kotsiantis, S. B., I. D. Zaharakis, and P. E. Pintelas. 2007. Supervised machine learning: A review of classification techniques. *Frontiers in Artificial Intelligence and Applications* 160: 3.

Lim, T., W. Loh, and Y. Shih. 2000. A comparison of prediction accuracy, complexity, and training time of thirty-three old and new classification algorithms. *Machine Learning* 40(3): 203–228.

Mitchell, T. M. 1997. *Machine Learning. 1997*, Burr Ridge, IL: McGraw Hill, p. 45.

Ng, R. T. et al. 1994. Efficient and effective clustering methods for spatial data mining. *20th Proceedings of the International Conference on Very Large Data Bases*, Santiago, Chile. pp. 144–155.

Vatsavai, R. R. et al. 2012. Spatiotemporal data mining in the era of big spatial data: algorithms and applications. *Proceedings of the 1st ACM SIGSPATIAL International Workshop on Analytics for Big Geospatial Data*. Redondo Beach, CA: ACM.

Wang, Z., Y. Song, and C. Zhang. 2009. Efficient active learning with boosting. *Proceedings of the SIAM Data Mining Conference (SDM 2009)*, Sparks, NV.

8 Spatial Big Data
Case Studies on Volume, Velocity, and Variety

Michael R. Evans, Dev Oliver, Xun Zhou, and Shashi Shekhar

CONTENTS

8.1 INTRODUCTION

Spatial computing encompasses the ideas, solutions, tools, technologies, and systems that transform our lives and society by creating a new understanding of spaces, their locations, places, and properties; how we know, communicate, and visualize our relation to places in a space of interest; and how we navigate through those places. From virtual globes to consumer global navigation satellite system devices, spatial computing is transforming society. With the rise of new spatial big data (SBD), spatial computing researchers will be working to develop a compelling array of new geo-related capabilities. We believe that these data, which we call SBD, represent the next frontier in spatial computing. Examples of emerging SBD include temporally detailed (TD) road maps that provide traffic speed values every minute for every road in a city, global positioning system (GPS) trajectory data from cell phones, engine measurements of fuel consumption, and greenhouse gas (GHG) emissions. A 2011 McKinsey Global Institute report defines traditional big data as data featuring one or more of the 3 *V's*: volume, velocity, and variety [1]. Spatial data frequently demonstrate at least one of these core features, given the variety of data types in spatial computing such as points, lines, and polygons. In addition, spatial analytics have shown to be more computationally expensive than their nonspatial brethren [2] as they need to account for spatial autocorrelation and nonstationarity, among other things.

In this chapter, we begin in Section 8.2 by defining SBD, enumerating three traditional categories of spatial data, and discussing their SBD equivalents. We then use case studies to demonstrate the 3 *V's* of SBD: volume, velocity, and variety [1]. A case study on climate data in Section 8.3 illustrates the challenges of utilizing large volumes of SBD. Velocity is demonstrated in Section 8.4 through a case study on loop detector (traffic speed) data on the Twin Cities, MN, highway network. Lastly, variety in SBD can refer to both the type of data input used and the variety in the type of output representations. We illustrate variety in data types through a case study on GPS trajectory data to find cyclist commuter corridors in Minneapolis, MN, and variety in data output is demonstrated through network activity summarization of pedestrian fatality data from Orlando, FL.

8.2 WHAT IS SPATIAL BIG DATA?

Spatial data are discrete representations of continuous phenomena. Discretization of continuous space is necessitated by the nature of digital representation. There are three basic models to represent spatial data: raster (grid), vector, and network. Satellite images are good examples of raster data. On the other hand, vector data consist of points, lines, polygons, and their aggregate (or multi-) counterparts. Graphs consisting of spatial networks are another important data type used to represent road networks. We define SBD as simply instances of these data types that exhibit at least one of the 3 *V's*: volume, velocity, and variety. In the following discussion, we provide examples of SBD in each of these core spatial data types: raster, vector, and network.

Raster data, such as geo-images (Google Earth), are frequently used for remote sensing and land classification. New spatial big raster datasets are emerging from a number of sources:

Unmanned aerial vehicle (UAV) data: Wide-area motion imagery sensors are increasingly being used for persistent surveillance of large areas, including densely populated urban areas. The wide-area video coverage and 24/7 persistent surveillance of these sensor systems allow for new and interesting patterns to be found via temporal aggregation of information. However, there are several challenges associated with using UAVs in gathering and managing raster datasets. First, UAV has a small footprint due to the relatively low flying height; therefore, it captures a large amount of images in a very short period of time to achieve the spatial coverage for many applications. This poses a significant challenge to store increasing large digital images. Image processing is another challenge because traditional approaches have shown to be too time-consuming and costly to rectify and mosaic the UAV photography for large areas. The large quantity of data far exceeds the capacity of the available pool of human analysts [3]. It is essential to develop automated, efficient, and accurate techniques to handle SBD as shown in Figure 8.1.

Light detection and ranging or laser imaging detection and ranging (Lidar): Lidar data are generated by timing laser pulses from an aerial position (plane or satellite) over a selected area to produce a surface mapping [5]. Lidar data are very rich for use cases related to surface analysis or feature extraction. However, these datasets are noisy and may contain irrelevant data for spatial analysis and sometimes miss critical information. These large volumes of data from multiple sources pose

(a)

(b)

FIGURE 8.1 SBD challenges through data volume, velocity, and variety. (a) Wide-area persistent surveillance. FOV: field of view. (Photo courtesy of the Defense Advanced Research Projects Agency (DARPA), Arlington, VA.) EO: electrooptical. (From Levchuk, G. et al., *Proc. SPIE*, 7704, 77040P, 2010.) (b) Lidar images of ground zero rendered September 27, 2001, by the US Army Joint Precision Strike Demonstration (JPSD) from data collected by NOAA flights. Thanks to NOAA/US Army JPSD.

a big challenge on management, analysis, and timely accessibility. Particularly, Lidar points and their attributes have tremendous sizes making them difficult to categorize these datasets for end users. Data integration from multiple spatial sources is another challenge due to the massive amounts of Lidar datasets. Therefore, SBD is an essential issue for Lidar remote sensing.

Vector data models over space are a framework to formalize specific relationships among a set of objects. Vector data consist of points, lines, and polygons, and with the rise of SBD, corresponding datasets have arisen from a variety of sources:

Volunteered geographic information (VGI) data: VGI brings a new notion of infrastructure to collect, synthesize, verify, and redistribute geographic data through geo-location technology, mobile devices, and geo-databases. These geographic data are provided, modified, and shared based on user interactive online services (e.g., OpenStreetMap, Wikimapia, Google Map, Google Earth, Microsoft's Virtual Earth, Flickr). In recent years, VGI leads an explosive growth in the availability of user-generated geographic information and requires scalable storage models to handle large-scale spatial datasets. The challenge for VGI is to enhance data service quality with regard to accuracy, credibility, reliability, and overall value [6].

GPS trace data: GPS trajectories are quickly becoming available for a larger collection of vehicles due to rapid proliferation of cell phones, in-vehicle navigation devices, and other GPS data-logging devices [7] such as those distributed by insurance companies [8]. Such GPS traces allow indirect estimation of fuel efficiency and GHG emissions via estimation of vehicle speed, idling, and congestion. They also make it possible to provide personalized route suggestions to users to reduce fuel consumption and GHG emissions. For example, Figure 8.2 shows 3 months of GPS trace data from a commuter with each point representing a GPS record taken at 1 min intervals, 24 h a day, 7 days a week. As can be seen, 3 alternative commute routes were identified between home and work from this dataset. These routes may be compared for engine idling that are represented by darker circles. Assuming the availability of a model to estimate fuel consumption from speed profiles, one may even rank alternative routes for fuel efficiency. In recent years, consumer GPS products [7,9] are evaluating the potential of this approach. Again, a key hurdle is the dataset size, which can reach 10^{13} items per year given constant minute-resolution measurements for all 100 million US vehicles.

Network data are commonly used to represent road maps for routing queries. While the network structure of the graph may not change, the amount of information about the network is rising drastically. New TD road maps give minute by minute speed information, along with elevation and engine measurements to allow for more sophisticated querying of road networks.

Spatiotemporal (ST) Engine Measurement Data: Many modern fleet vehicles include rich instrumentation such as GPS receivers, sensors to periodically measure subsystem properties [11–16], and auxiliary computing, storage, and communication devices to log and transfer accumulated datasets. Engine measurement datasets may be used to study the impacts of the environment (e.g., elevation changes, weather), vehicles (e.g., weight, engine size, energy source), traffic management systems (e.g., traffic light timing policies), and driver behaviors (e.g., gentle acceleration or braking) on fuel savings and GHG emissions. These datasets may include a time series of attributes such as vehicle location, fuel levels, vehicle speed, odometer values, engine speed in revolutions per minute (RPM), engine load, and emissions of GHGs (e.g., CO_2 and NO_X). Fuel efficiency can be estimated from fuel levels and distance traveled as well as engine idling from engine RPM. These attributes may be compared with geographic contexts such as elevation changes and traffic signal patterns to improve understanding of fuel efficiency and GHG emission. For example, Figure 8.3 shows heavy truck fuel consumption as a function of elevation from a recent study at Oak Ridge National Laboratory [17]. Notice how fuel consumption changes drastically with elevation slope changes. Fleet owners have studied such datasets to fine-tune routes to reduce unnecessary idling [18,19]. It is tantalizing to explore the potential of such datasets to help consumers

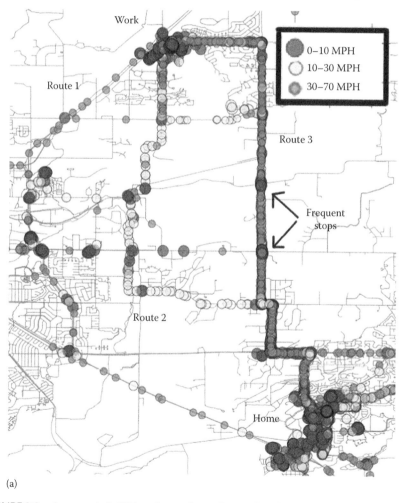

(a)

FIGURE 8.2 A commuter's GPS tracks over 3 months reveal preferred routes. (a) GPS trace data.

(*continued*)

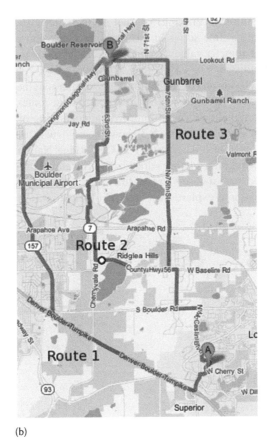

(b)

FIGURE 8.2 (continued)   A commuter's GPS tracks over 3 months reveal preferred routes. (b) Routes 1 through 3. (From Google Maps. http://maps.google.com.)

gain similar fuel savings and GHG emission reduction. However, these datasets can grow big. For example, measurements of 10 engine variables, once a minute, over the 100 million US vehicles in existence [20,21], may have 10^{14} data items per year.

Historical Speed Profiles: Typically, digital road maps consist of centerlines and topologies of road networks [22,23]. These maps are used by navigation devices

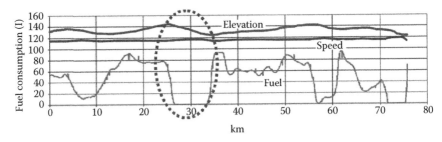

FIGURE 8.3   Engine measurement data improve understanding of fuel consumption. (From Capps G. et al., Class-8 heavy truck duty cycle project final report, ORNL/TM-2008/122, 2008.)

and web applications such as Google Maps [10] to suggest routes to users. New datasets from companies, such as NAVTEQ [24], use probe vehicles and highway sensors (e.g., loop detectors) to compile travel time information across road segments throughout the day and week at fine temporal resolutions (seconds or minutes). These data are applied to a profile model, and patterns in the road speeds are identified throughout the day. The profiles have data for every 5 min, which can then be applied to the road segment, building up an accurate picture of speeds based on historical data. Such TD road maps contain much more speed information than traditional road maps. While traditional road maps have only one scalar value of speed for a given road segment (e.g., Edge ID 1), TD road maps may potentially list speed/travel time for a road segment (e.g., EID 1) for thousands of time points; see Figure 8.4a, in a typical week. This allows a commuter to compare alternate start times in addition to alternative routes. It may even allow comparison of (start time, route) combinations

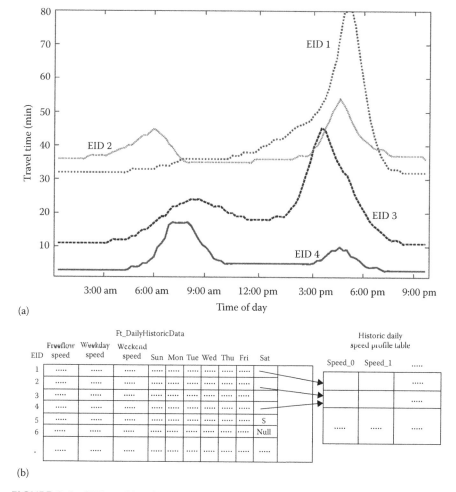

FIGURE 8.4 SBD on historical speed profiles. (a) Travel time along four road segments over a day. (b) Schema for daily historical speed data.

to select distinct preferred routes and distinct start times. For example, route ranking may differ across rush hour and non-rush hour and in general across different start times. However, TD road maps are large and their size may exceed 10^{13} items per year for the 100 million road segments in the United States when associated with per-minute values for speed or travel time. Thus, industry is using speed profiles, a lossy compression based on the idea of a typical day of a week, as illustrated in Figure 8.4b, where each (road segment, day of the week) pair is associated with a time series of speed values for each hour of the day.

8.3 VOLUME: DISCOVERING SUB-PATHS IN CLIMATE DATA

Sub-paths (i.e., intervals) in ST datasets can be defined as contiguous subsets of locations. Given an ST dataset and a path in its embedding ST framework, the goal of the interesting ST sub-path discovery problem is to identify all the dominant (i.e., not a subset of any other) interesting sub-paths along the path defined by a given interest measure. The ability to discover interesting sub-paths is important to many societal applications. For example, coastal area authorities may be interested in intervals of coastal lines that are prone to rapid environmental change due to rising ocean

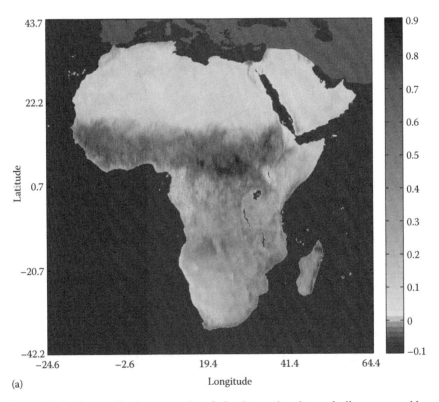

(a)

FIGURE 8.5 An application example of the interesting interval discovery problem. (a) Smoothed Africa vegetation dataset (measured in NDVI) in August 1981.

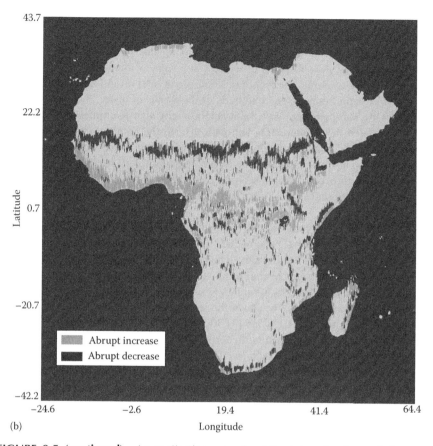

(b)   Longitude

FIGURE 8.5 (continued) An application example of the interesting interval discovery problem. (b) Longitudinal intervals of abrupt vegetation change in August 1981.

levels and melting polar icecaps. Water quality monitors may be interested in river segments where water quality changes abruptly.

An extended example from ecoclimate science illustrates the interesting sub-path discovery problem in details. This example comes from our collaboration with scientists studying the response of ecosystems to climate change by observing changes in vegetation cover across ecological zones. Sub-paths of abrupt vegetation cover change may serve to outline the spatial footprint of ecotones, the transitional areas between these zones [25]. Due to their vulnerability to climate changes, finding and tracking ecotones gives us important information about how the ecosystem responds to climate changes. Figure 8.5 illustrates the application of interesting sub-path discovery on the Africa vegetation cover in normalized difference vegetation index (NDVI) data, August 1981. Figure 8.5a shows a map of vegetation cover in Africa [26]. Each longitudinal path is taken as an input of the problem. The output, as shown in Figure 8.5b, is a map of longitudinal sub-paths with abrupt vegetation cover changes highlighted in dark and light shades of gray, shown in the legend. The footprints of several ecotones in Africa are discovered. One of them is the Sahel region, where vegetation cover exhibits an abrupt decreasing trend from south to north.

Discovering interesting sub-paths is challenging due to the following reasons. First, the length of the sub-paths of interest may vary, without a predefined maximum length. For example, the length of flood-prone interval in long rivers (e.g., the Gange, Mississippi) may extend hundreds or thousands of miles. Second, the interestingness in a sub-path may not exhibit monotonicity, that is, uninteresting intervals may be included in an interesting sub-path. Third, the data volume is potentially large. For example, consider the problem of finding all the interesting longitude sub-paths exhibiting abrupt change in an ecoclimate dataset with attributes such as vegetation, temperature, and precipitation over hundreds of years from different global climate models and sensor networks. The volume of such SBD ranges from terabytes to petabytes.

Previous works on interesting ST sub-path/interval discovery focused on change point detection using 1D or 2D approaches. The 1D approaches aim to find points in a time series where there is a shift in the data distribution [27–29]. Figure 8.6a shows a sample dataset in vegetation cover along a particular longitude. Figure 8.6b shows an sub-path in this dataset from location 5 to 11 whose data exhibits an abruptly increasing trend. In contrast, Figure 8.6c shows the output of a specific implementation of the popular statistical measure CUSUM [27,30] on the same data where only location 6 is identified as a point of interest (with abrupt change from below the mean to above the mean). The 2D approaches such as edge detection [31] aim at finding boundaries between different areas in an image. However, the footprints of an identified edge over each 1D path (e.g., row or column) are still points. The previous related works

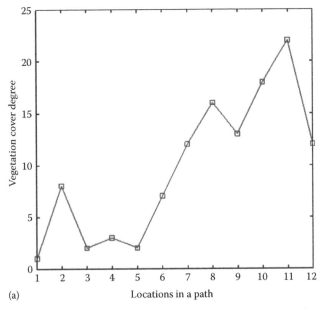

(a)

FIGURE 8.6 A comparison of interesting sub-paths in the data and change point found by related work. (a) Vegetation cover along a longitudinal path.

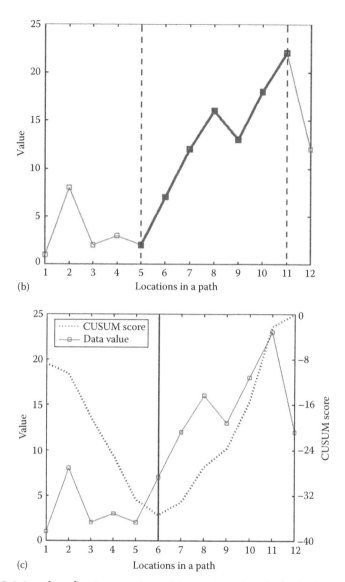

FIGURE 8.6 (continued) A comparison of interesting sub-paths in the data and change point found by related work. (b) An interesting sub-path in the path. (c) Result of CUSUM.

are limited to detecting points of interest in an ST path, rather than finding long interesting sub-paths/intervals. In contrast, our novel computational frameworks discover sub-paths of arbitrary length based on certain interest measures.

In our preliminary work [32], a sub-path enumeration and pruning (SEP) approach was proposed. In the approach, the enumeration space of all the sub-paths is modeled as a grid-based directed acyclic graph (G-DAG), where nodes are sub-paths and edges are subset relationships between sub-paths. The approach enumerates all the sub-path by performing a breadth-first traversal on the G-DAG, starting from the root (longest sub-path). Each sub-path is evaluated by commuting its algebraic interest measure. Should an interesting sub-path be identified, all its subsets are pruned. By doing this, we significantly reduce the number of sub-path evaluations. We apply this approach on ecoclimate datasets to find abrupt change sub-paths. An algebraic interest measure named *sameness degree* was designed to evaluate the change abruptness and persistence of a sub-path. As noted earlier, case study results on NDVI vegetation cover dataset showed that the approach can discover important patterns such as ecotones (e.g., the Sahel region). We also applied this approach on temporal paths (e.g., precipitation time series) and discovered patterns such as abrupt precipitation shifts in Africa. Experimental results on large synthetic datasets confirmed that the proposed approach is efficient and scalable.

8.4 VELOCITY: SPATIAL GRAPH OUTLIER DETECTION IN TRAFFIC DATA

In this section, we demonstrate SBD featuring velocity via a case study on real-time traffic monitoring datasets for detecting outliers in spatial graph datasets. We formalize the problem of ST outlier detection and propose an efficient graph-based outlier detection algorithm. We use our algorithm to detect spatial and temporal outliers in a real-world Minneapolis–St. Paul dataset and show effectiveness of our approach.

In 1997, the University of Minnesota and the Track Management Center Freeway Operations group started a joint project to archive sensor network measurements from the freeway system in the Twin Cities [33]. The sensor network includes about 900 stations, each of which contains one to four loop detectors, depending on the number of lanes. Sensors embedded in the freeways and interstate monitor the occupancy and volume of track on the road. At regular intervals, this information is sent to the track Management Center for operational purposes, for example, ramp meter control, as well as research on track modeling and experiments. Figure 8.7a shows a map of the stations on the highways within the Twin Cities metropolitan area, where each polygon represents one station. The interstate freeways include I-35W, I35E, I-94, I-394, I-494, and I-694. The state trunk highways include TH-100, TH-169, TH-212, TH-252, TH-5, TH-55, TH-62, TH-65, and TH-77. I-494 and I-694 together form a ring around the Twin Cities. I-94 passes from east to northwest, while I-35W and I-35E run in a south–north direction. Downtown Minneapolis is located at the intersection of I-94, I-394, and I-35W, and downtown Saint Paul is located at the intersection of I-35E and I-94. For each station, there is one detector installed in each lane. The track flow information measured by each detector can then be aggregated to the station level. The system records all the volume and occupancy information within each 5 min time slot at each particular station.

In this application, we are interested in discovering (1) the location of stations whose measurements are inconsistent with those of their graph-based spatial neighbors and (2) time periods when those abnormalities arise. We use three neighborhood definitions in this application as shown in Figure 8.7b. First, we define a neighborhood based on the spatial graph connectivity as a spatial graph neighborhood. In Figure 8.7b, $(s_1; t_2)$ and $(s_3; t_2)$ are the spatial neighbors of $(s_2; t_2)$ if s_1 and s_3 are connected to s_2 in a spatial graph. Second, we define a neighborhood based on a time series as a temporal neighborhood. In Figure 8.7b, $(s_2; t_1)$ and $(s_2; t_3)$ are the temporal neighbors of $(s_2; t_2)$ if t_1, t_2, and t_3 are consecutive time slots. In addition, we define a neighborhood based on both space and time series as a spatial–temporal neighborhood. In Figure 8.7b, $(s_1; t_1)$, $(s_1; t_2)$, $(s_1; t_3)$, $(s_2; t_1)$, $(s_2; t_3)$, $(s_3; t_1)$, $(s_3; t_2)$, and $(s_3; t_3)$ are the spatial–temporal neighbors of $(s_2; t_2)$ if s_1 and s_3 are connected to s_2 in a spatial graph and t_1, t_2, and t_3 are consecutive time slots.

The test for detecting an outlier can be described as follows: $\left| (S(x) - \mu_s)/\sigma_S \right| > \Theta$. For each data object x with an attribute value $f(x)$, the $S(x)$ is the difference of the attribute value of data object x and the average attribute value of its neighbors. μ_s is the mean value of all $S(x)$, and σ_s is the standard deviation of all $S(x)$. Choice of Θ depends on specified confidence interval. For example, a confidence interval of 95% will lead to $\Theta \approx 2$.

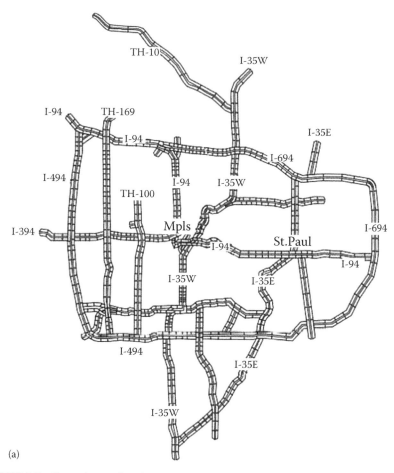

(a)

FIGURE 8.7 Detecting outliers in real-time traffic data. (a) Traffic speed detector map.

(continued)

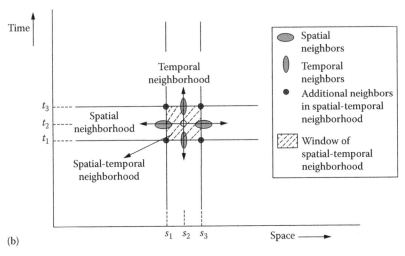

(b)

FIGURE 8.7 (continued)   Detecting outliers in real-time traffic data. (b) ST outlier detection framework.

In prior work [33], we proposed an I/O efficient algorithm to calculate the test parameters, for example, mean and standard deviation for the statistics. The computed mean and standard deviation can then be used to validate the outlier of the incoming dataset. Given an attribute dataset V and the connectivity graph G, the algorithm first retrieves the neighbor nodes from G for each data object x, and then it computes the difference of the attribute value of x to the average of the attribute values of x's neighbor nodes. These different values are then stored as a set. Finally, that set is computed to get the distribution value μ_s and σ_s. Note that the data objects are processed on a page basis to reduce redundant I/O. In other words, all the nodes within the same disk page are processed before retrieving the nodes of the next disk page.

The neighborhood aggregate statistics value, for example, mean and standard deviation, can be used to verify the outlier of an incoming dataset. The two verification procedures are route outlier detection and random node verification. The route outlier detection procedure detects the spatial outliers from a user-specified route. The random node verification procedure checks the outlierness from a set of randomly generated nodes. The step to detect outliers in both algorithms is similar, except that the random node verification has no shared data access needs across tests for different nodes. The storage of dataset should support I/O efficient computation of this operation.

Given a route RN in the dataset D with graph structure G, the route outlier detection algorithm first retrieves the neighboring nodes form G for each data object x in the route RN, and then it computes the difference $S(x)$ between the attribute value of x and the average of attribute values of x's neighboring nodes. Each $S(x)$ can then be tested using the spatial outlier detection test $\left| (S(x) - \mu_s)/\sigma_s \right| > \Theta$. The Θ is predetermined by the given confidence interval.

We tested the effectiveness of our algorithm on the Twin Cities track dataset and detected numerous outliers, as described in the following examples. In Figure 8.8b, the abnormal station (Station 139) was detected whose volume

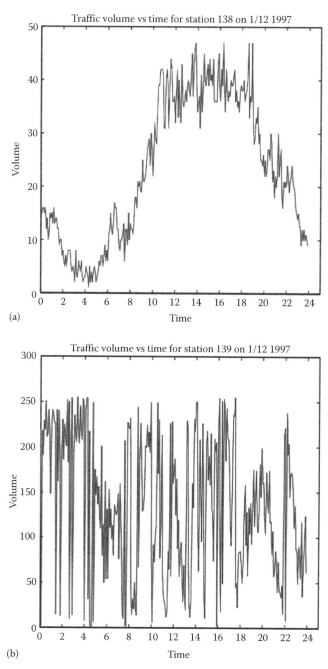

FIGURE 8.8 Outlier station 139 and its neighbor stations over 1 day. (a) Station 138. (b) Station 139.

(*continued*)

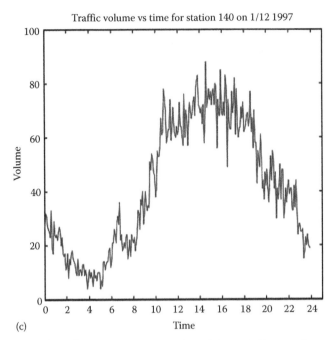

(c)

FIGURE 8.8 (continued) Outlier station 139 and its neighbor stations over 1 day. (c) Station 140.

values are significantly inconsistent with the volume values of its neighboring stations 138 and 140. Note that our basic algorithm detects outlier stations in each time slot; the detected outlier stations in each time slot are then aggregated to a daily basis.

Figure 8.8 shows an example of loop detector outliers. Figure 8.8a and c are the track volume maps for I-35W north bound and south bound, respectively, on January 21, 1997. The X-axis is the 5 min time slot for the whole day, and the Y-axis is the label of the stations installed on the highway, starting from 1 in the north end to 61 in the south end. The abnormal dark line at time slot 177 and the dark rectangle during time slot 100–120 on X-axis and between stations 29 and 34 on Y-axis can be easily observed from both Figure 8.8a and c. This dark line at time slot 177 is an instance of temporal outliers, where the dark rectangle is a spatial–temporal outlier. Moreover, station 9 in Figure 8.8a exhibits inconsistent track data compared with its neighboring stations and was detected as a spatial outlier.

8.5 VARIETY IN DATA TYPES: IDENTIFYING BIKE CORRIDORS

Given a set of trajectories on a road network, the goal of the k-primary corridor (k-PC) problem is to summarize trajectories into k groups, each represented by its most central trajectory. Figure 8.10a shows a real-world GPS dataset of a number of trips taken by bicyclists in Minneapolis, MN. The darkness indicates the usage levels of each road segment. The computational problem is summarizing this set of trajectories into a set of k-PCs. One potential solution is shown in Figure 8.9b for $k = 8$. Each identified primary corridor represents a subset of bike tracks from the original dataset. Note that the output of the k-PC problem is distinct from that of hot or frequent routes, as it is a summary of all the given trajectories partitioned into k-PCs.

The k-PC problem is important due to a number of societal applications, such as citywide bus route modification or bicycle corridor selection, among other urban

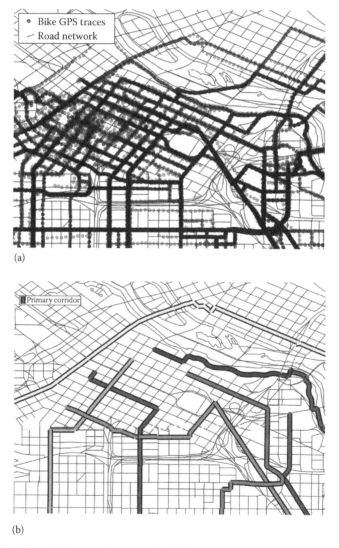

(a)

(b)

FIGURE 8.9 Example input and output of the k-PC problem. (a) Recorded GPS points from bicyclists in Minneapolis, MN. Intensity indicates number of points. (b) Set of 8 primary corridors identified from bike GPS traces in Minneapolis, MN.

development applications. Let us consider the problem of determining primary bicycle corridors through a city to facilitate safe and efficient bicycle travel. By selecting representative trajectories for a given group of commuters, the overall alteration to commuters routes is minimized, encouraging use. Facilitating commuter bicycle traffic has shown in the past to have numerous societal benefits, such as reduced GHG emissions and healthcare costs [34].

Clustering trajectories on road networks is challenging due to the computational cost of computing pairwise graph-based minimum-node-distance similarity metrics (e.g., network Hausdorff distance [NHD]) between trajectories in large GPS datasets as shown in our previous work [35]. We proposed a baseline algorithm using a graph-based approach to compute a single element of the TSM, requiring multiple invocations of common shortest-path algorithms (e.g., Dijkstra [36]). For example, given two trajectories consisting of 100 nodes each, a baseline approach to calculate NHD would need to compute the shortest distance between all pairs of nodes (10^4), which over a large trajectory dataset (e.g., 10,000 trajectories) would require 10^{12} shortest-path distance computations. This quickly becomes computationally prohibitive without faster algorithms.

Trajectory pattern mining is a popular field with a number of interesting problems both in geometric (Euclidean) spaces [37] and networks (graphs) [38]. A key component to traditional data mining in these domains is the notion of a similarity metric, the measure of sameness or closeness between a pair of objects. A variety of trajectory similarity metrics, both geometric and network, have been proposed in the literature [39]. One popular metric is Hausdorff distance, a commonly used measure to compare similarity between two geometric objects (e.g., polygons, lines, sets of points) [40]. A number of methods have focused on applying Hausdorff distance to trajectories in geometric space [41–44].

We formalize the NHD and propose a novel approach, that is, orders of magnitude faster than the baseline approach and our previous work, allowing NHD to be computed efficiently on large trajectory datasets. A baseline approach for solving the k-PC problem would involve comparing all nodes in each pairwise combination of trajectories. That is, when comparing two trajectories, each node in the first trajectory needs to find the shortest distance to any node in the opposite trajectory. The maximum value found is the Hausdorff distance [40]. While this approach does find the correct Hausdorff distance between the two trajectories, it is computationally expensive due to the node-to-node pairwise distance computations between each pair of trajectories. We will demonstrate using experimental and analytical analysis how prohibitive that cost is on large datasets. Due to this, related work solving the k-PC problem has resulted in various heuristics [45–49].

We propose a novel approach that ensures correctness while remaining computationally efficient. While the baseline approach depends on computing node-to-node distances, NHD requires node-to-trajectory minimum distance. We take advantage of this insight to compute the NHD between nodes and trajectories directly by modifying the underlying graph and computing from a trajectory to an entire set of trajectories with a single shortest-path distance computation. This approach retains correctness while proving significantly faster than the baseline approach and our previous work [35].

Our previous work [35] in the k-PC problem requires $O(|T|^2)$ invocations of a shortest-path algorithm to compute the necessary TSM, becoming prohibitively expensive when dealing with datasets with a large number of trajectories. Therefore, we developed a novel row-wise algorithm to compute the k-PC problem on SBD.

Computing $NHD(t_x,t_y)$ between two trajectories does not require the shortest distance between all pairs of nodes in t_x and t_y. We require the shortest network distance from each node in t_x to the *closest* node in t_y. In [35], we proposed a novel approach to find this distance, as compared to enumerating the all-pair shortest-path distances as the baseline approach does. This significantly reduced the number of distance calculations and node iterations needed to compute the TSM. In Figure 8.10b, to calculate $NHD(t_B,t_A)$, we began by inserting a virtual node ($A_{virtual}$) representing trajectory t_A into the graph. This node had edges with weights of 0 connecting it to each other node in trajectory t_A. We then ran a shortest-path distance computation from the virtual node as a source, with the

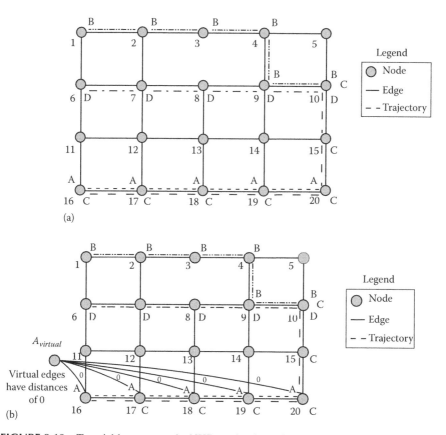

FIGURE 8.10 To quickly compute the NHD on the shown input (a), we alter the underlying graph to quickly compute distances to all nodes in the relevant tracks (b). (a) Road network represented as an undirected graph with four trajectories illustrated with bold dashed lines. (b) Inserting a virtual node ($A_{virtual}$) to represent track A for efficient NHD computation

destination being every node in trajectory t_B. The result was the shortest distance from each node in trajectory t_B to the virtual node $A_{virtual}$. Since the virtual node is only connected to nodes in trajectory t_A and all the edge weights are 0, we had the shortest path from each node in trajectory t_B to the closest node in trajectory t_A, exactly what $NHD(t_B,t_A)$ requires for computation. However, our previous work [35] focused on computing a single cell in the TSM per invocation of a single-source shortest-path algorithm. That meant at least $O(|T|^2)$ shortest-path invocations to compute the TSM for the k-PC problem, still quite expensive. We propose a new algorithm, row-based track similarity (ROW-TS), to compute an entire *row* of the TSM with one invocation of a single-source shortest-path algorithm. Using a row-based approach, we can essentially calculate $NHD(t \in T, t_A)$ with one single-source shortest-path invocation from $A_{virtual}$. This approach reduces the overall number of shortest-path invocations to $O(|T|)$ at the cost of additional bookkeeping, as we will show in the following discussion. However, due to the expensive cost of shortest-path algorithms, this results in significant performance savings.

In summer 2006, University of Minnesota researchers collected a variety of data to help gain a better understanding of commuter bicyclist behavior using GPS equipment and personal surveys to record bicyclist movements and behaviors [50]. The broad study examined a number of interesting issues with commuter cycling, for example, results showed that as perceived safety decreases (possibly due to nearby vehicles), riders appear to be more cautious and move more slowly. One possible issue they looked at was identifying popular transportation corridors for the Minnesota Department of Transportation to focus funds and (on?) repairs. At the time, they handcrafted the primary corridors. Shortly after this study, the US Department of Transportation (DOT) began a 4-year, $100 million pilot project in four communities (including Minneapolis) aimed to determine whether investing in bike and pedestrian infrastructure encouraged significant increases in public use. As a direct result of this project, the US DOT found that biking increased 50%, 7700 fewer tons of carbon dioxide was emitted, 1.2 fewer gallons of gas was burned, and there was a $6.9 million/year reduction in healthcare costs [34].

8.6   VARIETY IN OUTPUT: SPATIAL NETWORK ACTIVITY SUMMARIZATION

Spatial network activity summarization (SNAS) is important in several application domains including crime analysis and disaster response [51]. For example, crime analysts look for concentrations of individual events that might indicate a series of related crimes [52]. Crime analysts need to summarize such incidents on a map, so that law enforcement is better equipped to make resource allocation decisions [52]. In disaster response-related applications, action is taken immediately after a disastrous event with the aim of saving life; protecting property; and dealing with immediate disruption, damage, or other effects caused by the disaster [53]. Disaster response played an important role in the 2010 earthquake in Haiti, where there were many requests for assistance such as food, water, and medical supplies [54]. Emergency managers need the means to summarize these requests efficiently so that they can better understand how to allocate relief supplies.

The SNAS problem is defined informally as follows. Given a spatial network, a collection of activities and their locations (e.g., a node or an edge), a set of paths P, and a desired number of paths k, find a subset of k paths in P that maximizes the number of activities on each path. An activity is an object of interest in the network. In crime analysis, an activity could be the location of a crime (e.g., theft). In disaster response-related applications, an activity might be the location of a request for relief supplies. Figures 8.10a and 8.9b illustrate an input and output example of SNAS, respectively. The input consists of 6 nodes, 8 edges (with edge weights of one for simplicity), 14 activities, the set of shortest paths for this network, and $k = 2$, indicating that two routes are desired. The output contains two routes from the given set of shortest paths that maximize the activity coverage; route <C,A,B> covers activities 1 through 5 and route <C,D,E> covers activities 7, 8, and 12 through 14.

In network-based summarization, spatial objects are grouped using network (e.g., road) distance. Existing methods of network-based summarization such as mean streets [55], maximal subgraph finding (MSGF) [56], and clumping [57–64] group activities over multiple paths, a single path/subgraph, or no paths at all. Mean streets [55] find anomalous streets or routes with unusually high activity levels. It is not designed to summarize activities over k paths because the number of high-crime streets returned is always relatively small. MSGF [56] identifies the maximal subgraph (e.g., a single path, $k = 1$) under the constraint of a user-specified length and cannot summarize activities when $k > 1$. The network-based variable-distance clumping method (NT-VCM) [64] is an example of the clumping technique [57–64]. NT-VCM groups activities that are within a certain shortest-path distance of each other on the network; in order to run NT-VCM, a distance threshold is needed.

In this chapter, we propose a K-Main Routes (KMR) approach that finds a set of k routes to summarize activities. KMR aims to maximize the number of activities covered on each k route. KMR employs an inactive node pruning algorithm where instead of calculating the shortest paths between all pair of nodes, only the shortest paths between active nodes and all other nodes in the spatial network are calculated. This results in computational savings (without affecting the resulting summary paths) that are reported in the experimental evaluation. The inputs of KMR include (1) an undirected spatial network $G = (N,E)$; (2) a set of activities A; and (3) a number of routes, k, where $k \geq 1$. The output of KMR is a set of k routes where the objective is to maximize the activity coverage of each k route. Each k route is a shortest path between its end nodes, and each activity $a_i \in A$ is associated with only one edge $e_i \in E$.

KMR first calculates the shortest paths between active nodes and all other nodes and then selects k shortest paths as initial summary paths. The main loop of KMR (phases 1 and 2) is repeated until the summary paths do not change.

Phase 1: Assign activities to summary paths. The first step of this phase initializes the set of next clusters, that is, *nextClusters*, to the empty set. In general, this phase is concerned with forming k clusters by assigning each activity to its nearest summary path. To accomplish this, KMR considers each activity and each cluster in determining the nearest summary path to an activity.

KMR uses the following proximity measure to quantify nearness: $prox(s_i,a_i)$. The distance between an activity and a summary path is the minimum network distance between each node of the edge that the activity is on and each node of the summary path. Once the distance from each activity to each summary path is calculated, the activity is assigned to the cluster with the nearest summary path.

Phase 2: Recompute summary paths. This phase is concerned with recomputing the summary path of each cluster so as to further maximize the activity coverage. This entails iterating over each cluster and initializing the summary path for each cluster. The summary path of each cluster is updated based on the activities assigned to the cluster. The summary path with the maximum activity coverage is chosen as the new summary path for each cluster c_i, that is, $sp_{max} \leftarrow max(AC(sp_k) \in c_i \mid k=1...|sp|)$. Phases 1 and 2 are repeated until the summary paths of each cluster do not change. At the end of each iteration the current clusters are initialized to the next clusters. Once the summary paths of each cluster do not change, a set of k routes is returned.

We conducted a qualitative evaluation of KMR comparing its output with the output of CrimeStat [65] K-means [66] (a popular summarization technique) on a real pedestrian fatality dataset [67], shown in Figure 8.11a. The input consists of 43 pedestrian fatalities (represented as dots) in Orlando, Florida, occurring between 2000 and 2009. As we have explained, KMR uses paths and network distance to group activities on a spatial network. By contrast, in geometry-based summarization, the partitioning of spatial data is based on grouping similar points distributed in planar space where the distance is calculated using Euclidean distance. Such techniques focus on the discovery of the geometry (e.g., circle, ellipse) of high-density regions [52] and include K-means [66,68–71], K-medoid [72,73], P-median [74], and nearest neighbor hierarchical clustering [75] algorithms.

Figure 8.11b through d shows the results of KMR, K-means using Euclidean distance, and K-means using network distance, respectively. In all cases, K was set to 4. The output of each technique shows (1) the partitioning of activities represented by different shades of gray and (2) the representative of each partition (e.g., paths or ellipses). For example, Figure 8.11c shows (1) activities that are shaded respective to the four different partitions to which each activity belongs, and (2) ellipses representing each partition of activities.

This work explored the problem of SNAS in relation to important application domains such as crime analysis and disaster response. We proposed a KMR algorithm that discovers a set of k paths to group activities. KMR uses inactive node pruning, network Voronoi activity assignment, and divide and conquer summary path recomputation to enhance its performance and scalability. Experimental evaluation using both synthetic and real-world datasets indicated that the performance-tuning decisions utilized by KMR yielded substantial computational savings without reducing the coverage of the resulting summary paths. For qualitative evaluation, a case study comparing the output of KMR with the output of a current geometry-based summarization technique highlighted the potential usefulness of KMR to summarize activities on spatial networks.

8.7 SUMMARY

Increasingly, location-aware datasets are of a size, variety, and update rate that exceed the capability of spatial computing technologies. This chapter discussed some of the emerging challenges posed by such datasets, referred to as SBD. SBD examples include trajectories of cell phones and GPS devices, TD road maps, and vehicle engine measurements. SBD has the potential to transform society. A recent McKinsey Global Institute report estimates that SBD, such as personal location data, could save consumers hundreds of billions of dollars annually by 2020 by helping vehicles avoid congestion via next-generation routing services such as eco-routing.

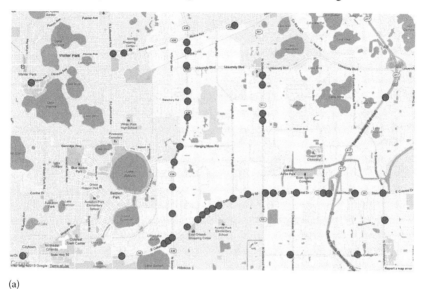

(a)

(b)

FIGURE 8.11 Comparing KMR and CrimeStat *K*-means output for *k* = 4 on pedestrian fatality data from Orlando, FL. (From Fatality Analysis Reporting System (FARS), National highway traffic safety administration (NHTSA), http://www.nhtsa.gov/FARS.) (a) Input. (b) KMR.

(continued)

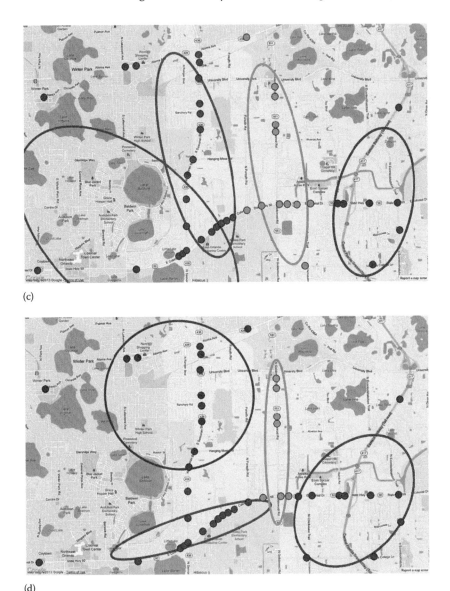

(c)

(d)

FIGURE 8.11 (continued) Comparing KMR and CrimeStat K-means output for $k = 4$ on pedestrian fatality data from Orlando, FL. (c) CrimeStat K-means with Euclidean distance. (d) CrimeStat K-means with network distance.

SBD has immense potential to benefit a number of societal applications. By harnessing this increasingly large, varied, and changing data, new opportunities to solve worldwide problems are presented. To capitalize on these new datasets, inherent challenges that come with SBD need to be addressed. For example, many spatial operations are iterative by nature, something that parallelization has not yet been able to handle completely. By expanding cyber-infrastructure, we can harness the power of these massive spatial datasets. New forms of analytics using simpler models and richer neighborhoods will enable solutions in a variety of disciplines.

REFERENCES

1. J. Manyika et al., Big data: The next frontier for innovation, competition, and productivity, McKinsey Global Institute, 2011.
2. S. Shekhar, M. Evans, J. Kang, and P. Mohan, Identifying patterns in spatial information: A survey of methods, *Wiley Interdisciplinary Reviews: Data Mining and Knowledge Discovery*, 1(3), 193–214, 2011. Wiley Online Library.
3. New York Times, Military is awash in data from drones. http://www.nytimes.com/2010/01/11/business/11drone.html?pagewanted = all, 2010 (Accessed: July 1, 2013).
4. G. Levchuk, A. Bobick, and E. Jones, Activity and function recognition for moving and static objects in urban environments from wide-area persistent surveillance inputs, *Proc. SPIE 7704, Evolutionary and Bio-Inspired Computation: Theory and Applications IV*, 77040P, April 15, 2010, doi:10.1117/12.849492.
5. New York Times, Mapping ancient civilization, in a matter of days. http://www.nytimes.com/2010/05/11/science/11maya.html, 2010 (Accessed: July 1, 2013).
6. InformationWeek, Red cross unveils social media monitoring operation. http://www.informationweek.com/government/information-management/red-cross-unveils-social-media-monitorin/232602219, 2012.
7. Garmin Ltd. Garmin. http://www.garmin.com/us/ (Accessed: July 1, 2013).
8. P. De Corla-Souza, Estimating benefits from mileage-based vehicle insurance, taxes, and fees. *Transportation Research Record: Journal of the Transportation Research Board* 1812(1), 171–178, 2002.
9. TomTom, TomTom GPS navigation. http://www.tomtom.com/ (Accessed: October 29, 2011).
10. Google. Google maps. https://maps.google.com (Accessed: July 1, 2013).
11. H. Kargupta, J. Gama, and W. Fan, The next generation of transportation systems, greenhouse emissions, and data mining, in *Proceedings of the 16th ACM SIGKDD International Conference on Knowledge Discovery and Data Mining*, pp. 1209–1212, ACM, New York, 2010.
12. H. Kargupta, V. Puttagunta, M. Klein, and K. Sarkar, On-board vehicle data stream monitoring using minefleet and fast resource constrained monitoring of correlation matrices, *New Generation Computing*, 25(1), 5–32, 2006. Springer.
13. Lynx Technologies. Lynx GIS. http://www.lynxgix.com (Accessed: July 1, 2013).
14. MasterNaut, Green solutions. http://www.masternaut.co.uk/carbon-calculator/ (Accessed: July 1, 2013).
15. Telenav, Inc. Telenav. http://www.telenav.com (Accessed: July 1, 2013).
16. TeloGIS. TeloGIS. http://www.telogis.com (Accessed: July 1, 2013).
17. G. Capps, O. Franzese, B. Knee, M. Lascurain, and P. Otaduy, Class-8 heavy truck duty cycle project final report, ORNL/TM-2008/122, 2008. Oak Ridge National Laboratory, Oakridge, TN.
18. A. T. R. I. (ATRI), Fpm congestion monitoring at 250 freight significant highway location: Final results of the 2010 performance assessment. http://goo.gl/3cAjr (Accessed: July 1, 2013), 2010.
19. A. T. R. I. (ATRI), Atri and fhwa release bottleneck analysis of 100 freight significant highway locations. http://goo.gl/C0NuD (Accessed: July 1, 2013), 2010.
20. D. Sperling and D. Gordon, *Two Billion Cars*. Oxford University Press, Oxford, U.K., 2009.

21. Federal Highway Administration, *Highway Statistics*, HM-63, HM-64, 2008.
22. B. George and S. Shekhar, Road maps, digital, in *Encyclopedia of GIS*, pp. 967–972, Springer, 2008.
23. S. Shekhar and H. Xiong, *Encyclopedia of GIS*. Springer Publishing Company, Incorporated, New York, 2007.
24. NAVTEQ. NAVTEQ. http://www.navteq.com (Accessed: July 1, 2013).
25. I. Noble, A model of the responses of ecotones to climate change, *Ecological Applications*, 3(3), 396–403, 1993. JSTOR.
26. C.J. Tucker, J.E. Pinzon, and M.E. Brown, Global inventory modeling and mapping studies. *Global Land Cover Facility*, University of Maryland, College Park, MD, 1981–2006.
27. E.S. Page, Continuous inspection schemes, *Biometrika*, 41(1/2), 100–115, 1954. JSTOR.
28. D. Nikovski and A. Jain, Fast adaptive algorithms for abrupt change detection, *Machine Learning*, 79(3), 283–306, 2010. Springer.
29. M. Sharifzadeh, F. Azmoodeh, and C. Shahabi, Change detection in time series data using wavelet footprints, *Advances in Spatial and Temporal Databases*, pp. 127–144, 2005, Springer, Berlin, Germany.
30. J. Kucera, P. Barbosa, and P. Strobl, Cumulative sum charts-a novel technique for processing daily time series of modis data for burnt area mapping in portugal, in *International Workshop on the Analysis of Multi-temporal Remote Sensing Images*, pp. 1–6, IEEE, Piscataway, NJ, 2007.
31. J. Canny, A computational approach to edge detection, *Readings in Computer Vision: Issues, Problems, Principles, and Paradigms*, vol. 184, no. 87–116, p. 86, 1987, Morgan Kaufmann, San Francisco, CA.
32. X. Zhou, S. Shekhar, P. Mohan, S. Liess, and P. Snyder, Discovering interesting subpaths in spatiotemporal datasets: A summary of results, in *Proceedings of the 19th International Conference on Advances in Geographical Information Systems (ACMGIS 2011)*, ACM, Chicago, IL, November 1–4, 2011.
33. S. Shekhar, C.-T. Lu, and P. Zhang, Detecting graph-based spatial outliers: Algorithms and applications (a summary of results), in *Proceedings of the Seventh ACM SIGKDD International Conference on Knowledge Discovery and Data Mining*, pp. 371–376, ACM, New York, 2001.
34. J. Marcotty, Federal funding for bike routes pays off in twin cities. http://www.startribune.com/local/minneapolis/150105625.html (Accessed: July 1, 2013).
35. M.R. Evans, D. Oliver, S. Shekhar, and F. Harvey, Summarizing trajectories into k-primary corridors: A summary of results, in *Proceedings of the 20th International Conference on Advances in Geographic Information Systems*, (*SIGSPATIAL '12*), pp. 454–457, ACM, New York, 2012.
36. T. Cormen, *Introduction to Algorithms*. The MIT press, Cambridge, U.K., 2001.
37. Y. Zheng and X. Zhou, *Computing with Spatial Trajectories*, 1st edn. Springer Publishing Company, Incorporated, New York, 2011.
38. R.H. Güting, V.T. De Almeida, and Z. Ding, Modeling and querying moving objects in networks, *The VLDB Journal*, 15(2), 165–190, 2006.
39. B. Morris and M. Trivedi, Learning trajectory patterns by clustering: Experimental studies and comparative evaluation, in *IEEE Conference on Computer Vision and Pattern Recognition*, (*CVPR 2009*), pp. 312–319, June 20–25, 2009.
40. D.P. Huttenlocher, K. Kedem, and J.M. Kleinberg, On dynamic Voronoi diagrams and the minimum Hausdorff distance for point sets under Euclidean motion in the plane, in *Proceedings of the Eighth Annual Symposium on Computational Geometry (SCG'92)*, pp. 110–119, ACM, New York, NY, 1992.
41. J. Henrikson, Completeness and total boundedness of the Hausdorff metric, *MIT Undergraduate Journal of Mathematics*, 1, 69–80, 1999.

42. S. Nutanong, E.H. Jacox, and H. Samet, An incremental Hausdorff distance calculation algorithm, in *Proceedings of the VLDB Endowment*, vol. 4, no. 8, pp. 506–517, May 2011.

43. J. Chen, R. Wang, L. Liu, and J. Song, Clustering of trajectories based on Hausdorff distance, in *International Conference on Electronics, Communications and Control (ICECC)*, pp. 1940–1944, IEEE, September 9–11, 2011.

44. H. Cao and O. Wolfson, Nonmaterialized motion information in transport networks, In *Proceedings of the 10th International Conference on Database Theory (ICDT'05)*, T. Eiter and L. Libkin (Eds.), pp. 173–188, Springer-Verlag, Berlin, Germany, 2005.

45. G. Roh and S. Hwang, Nncluster: An efficient clustering algorithm for road network trajectories, in *Database Systems for Advanced Applications*, pp. 47–61, Springer, Berlin, Germany, 2010.

46. J.-R. Hwang, H.-Y. Kang, and K.-J. Li, Spatio-temporal similarity analysis between trajectories on road networks, in *Proceedings of the 24th International Conference on Perspectives in Conceptual Modeling*, ER'05, pp. 280–289, Springer-Verlag, Berlin, Germany, 2005.

47. J.-R. Hwang, H.-Y. Kang, and K.-J. Li, Searching for similar trajectories on road networks using spatio-temporal similarity, in *Advances in Databases and Information Systems*, pp. 282–295, Springer, Berlin, Germany, 2006.

48. E. Tiakas, A.N. Papadopoulos, A. Nanopoulos, Y. Manolopoulos, D. Stojanovic, and S. Djordjevic-Kajan, Searching for similar trajectories in spatial networks, *Journal of Systems and Software*, 82, 772–788, May 2009.

49. E. Tiakas, A.N. Papadopoulos, A. Nanopoulos, Y. Manolopoulos, D. Stojanovic, and S. Djordjevic-Kajan, Trajectory similarity search in spatial networks, in *10th International Database Engineering and Applications Symposium, 2006 (IDEAS'06)*, pp. 185–192, IEEE, Delhi, India, 2006.

50. F. Harvey and K. Krizek, *Commuter Bicyclist Behavior and Facility Disruption*, Technical Report no. MnDOT 2007-15, University of Minnesota, Minneapolis, MN, 2007.

51. D. Oliver, A. Bannur, J.M. Kang, S. Shekhar, and R. Bousselaire, A k-main routes approach to spatial network activity summarization: A summary of results, in *ICDM Workshops*, pp. 265–272, Sydney, New South Wales, Australia, 2010.

52. J. Eck, S. Chainey, J. Cameron, M. Leitner, and R. Wilson, Mapping crime: Understanding hot spots, 2005. National Institute of Justice.

53. W. Carter, *Disaster Management: A Disaster Manager's Handbook*. Asian Development Bank, Manila, Philippines, 1991.

54. Ushahidi. Crisis map of Haiti. http://haiti.ushahidi.com (Accessed: July 1, 2013).

55. M. Celik, S. Shekhar, B. George, J. Rogers, and J. Shine, *Discovering and Quantifying Mean Streets: A Summary of Results*, Technical Report 07-025, University of Minnesota, Computer Science and Engineering, Minneapolis, MN, 2007.

56. K. Buchin, S. Cabello, J. Gudmundsson, M. Löffler, J. Luo, G. Rote, R.I. Silveira, B. Speckmann, and T. Wolle, Detecting hotspots in geographic networks, *Advances in GIScience*, pp. 217–231, Springer, Berlin, Germany, 2009.

57. S. Roach, *The Theory of Random Clumping*. Methuen, London, U.K., 1968.

58. A. Okabe, K. Okunuki, and S. Shiode, The SANET toolbox: New methods for network spatial analysis, *Transactions in GIS*, 10(4), 535–550, 2006.

59. S. Shiode and A. Okabe, Network variable clumping method for analyzing point patterns on a network, in *Unpublished Paper Presented at the Annual Meeting of the Associations of American Geographers*, Philadelphia, PA, 2004.

60. K. Aerts, C. Lathuy, T. Steenberghen, and I. Thomas, Spatial clustering of traffic accidents using distances along the network, in *11th World Conference on Transport Research*, 2007, Berkley, CA.

61. P. Spooner, I. Lunt, A. Okabe, and S. Shiode, Spatial analysis of roadside *Acacia* populations on a road network using the network K-function, *Landscape Ecology*, 19(5), 491–499, 2004.

62. T. Steenberghen, T. Dufays, I. Thomas, and B. Flahaut, Intra-urban location and clustering of road accidents using GIS: A Belgian example, *International Journal of Geographical Information Science*, 18(2), 169–181, 2004.

63. I. Yamada and J. Thill, Local indicators of network-constrained clusters in spatial point patterns, *Geographical Analysis*, 39(3), 268–292, 2007.

64. S. Shiode and N. Shiode, Detection of multi-scale clusters in network space, *International Journal of Geographical Information Science*, 23(1), 75–92, 2009.

65. N. Levine, *CrimeStat: A Spatial Statistics Program for the Analysis of Crime Incident Locations (v 2.0)*, Ned Levine & Associates, Houston, TX, and the National Institute of Justice, Washington, DC, 2002.

66. J. MacQueen et al., Some methods for classification and analysis of multivariate observations, in *Proceedings of the Fifth Berkeley Symposium on Mathematical Statistics and Probability*, vol. 1, p. 14, Berkeley, CA, 1967.

67. Fatality Analysis Reporting System (FARS), National highway traffic safety administration (NHTSA), http://www.nhtsa.gov/FARS.

68. S. Borah and M. Ghose, Performance analysis of AIM-K-means & K-means in quality cluster generation, *Journal of Computing*, 1(1), 175–178, 2009.

69. A. Barakbah and Y. Kiyoki, A pillar algorithm for K-Means optimization by distance maximization for initial centroid designation, *IEEE Symposium on Computational Intelligence and Data Mining (CIDM)*, Nashville, TN, 2009.

70. S. Khan and A. Ahmad, Cluster center initialization algorithm for K-means clustering, *Pattern Recognition Letters*, 25(11), 1293–1302, 2004.

71. D. Pelleg and A. Moore, X-means: Extending K-means with efficient estimation of the number of clusters, in *Proceedings of the Seventeenth International Conference on Machine Learning*, pp. 727–734, San Francisco, CA, 2000.

72. L. Kaufman and P. Rousseeuw, *Finding Groups in Data: An Introduction to Cluster Analysis*. Wiley Online Library, New York, 1990.

73. R.T. Ng and J. Han, Efficient and effective clustering methods for spatial data mining, in *Proceedings of the 20th International Conference on Very Large Data Bases (VLDB'94)*, J.B. Bocca, M. Jarke, and C. Zaniolo (Eds.), Morgan Kaufmann Publishers Inc., pp. 144–145, San Francisco, CA, 1994.

74. M. Resende and R. Werneck, A hybrid heuristic for the p-median problem, *Journal of Heuristics*, 10(1), 59–88, 2004.

75. R. D'Andrade, U-statistic hierarchical clustering, *Psychometrika*, 43(1), 59–67, 1978.

9 Exploiting Big VGI to Improve Routing and Navigation Services

Mohamed Bakillah, Johannes Lauer,
Steve H.L. Liang, Alexander Zipf,
Jamal Jokar Arsanjani, Amin Mobasheri,
and Lukas Loos

CONTENTS

9.1　INTRODUCTION

The recent technological advances in data production and dissemination have enabled the generation of unprecedented volumes of geospatial data, giving rise to the paradigm of big data. Part of this trend is volunteered geographic information (VGI), that is, geographic information (GI) produced by volunteer contributors (Goodchild 2007), and crowdsourced data such as those obtained from social media. Whether *big data* refer to these huge volumes of data, which no longer fit traditional database structures (Dumbill 2013), or to the new technologies and techniques that must be developed to deal with these massive data sets (Davenport et al. 2012), there is a common understanding that the full potential of such amount of data can only be exploited if information and knowledge with added value can be extracted from it.

As for many applications, the possibility to exploit massive data sets raises new opportunities to improve mobility services, including routing and navigation services. Traditionally, routing and navigation services relied on the digitization of the existing streets by commercial mapping agencies and mainly two companies (NAVTEQ and TomTom [formerly TeleAtlas]). The emergence of VGI and crowdsourcing-based applications is changing this perspective by offering a wide range of additional information that can be used to complement the official data produced by governments and mapping agencies. VGI and crowdsourcing applications have the potential to offer more diverse, more detailed, more local, and more contextualized data. Data from VGI and crowdsourcing applications with a potential of leveraging billion of contributors can fulfil the gaps found in official data that cannot be updated quickly at a low cost. For instance, if VGI and crowdsourcing applications could be exploited into some extent, developers could think of developing routing and navigation services that recommended routes personalized according to the travelers' context. However, near real-time interaction with the changes detected on the road network through various sources could help reduce congestion, energy consumption, greenhouse gas and particle emissions, etc.

While VGI and crowdsourcing applications are clearly a sort of big geo-data, they feature some particularities that make their exploitation challenging at the moment. This chapter is dedicated to the analysis of the potential of VGI and crowdsourced data for improving routing and navigation services. The limitations of these types of data and how their exploitation is challenging are also discussed. Based on these limitations, we suggest some avenues for future research on the next generation of collaborative routing and navigation services.

This chapter begins with briefly reviewing the paradigms of big data and VGI. By providing a background to traditional routing and navigation services, we explain the types of big VGI data sources that could be exploited to upgrade routing and navigation services. Then, drawing from the limitations of VGI and crowdsourced data, we discuss the challenges for exploiting big VGI to improve routing and navigation services.

9.2   WHAT IS BIG DATA?

Recent technological advances in data production and dissemination have enabled the generation of unprecedented volumes of geospatial data. In 2012, it was estimated that the global volume of data was growing at a 50% rate each year (Lohr 2012), due to the increasing dissemination of digital sensors, smart phones, GPS-enabled devices, crowdsourcing applications, and social media, among other phenomena. While geospatial data have traditionally remained at the hand of experts (governments, mapping agencies), paradigms such as open data, social media, and collaborative mapping projects make it possible for an increasing proportion of these data to be virtually available to anyone, with the potential to benefit businesses, civil society, and individuals in general.

This huge amount of data has given rise to the term *big data*. Big data is a loosely used term that is employed to refer to two ideas. First, it is used to refer to the huge volume of data itself. For example, Dumbill (2013) states that "Big Data is data that exceeds the processing capacity of conventional data systems" and does not "fit the structure of your database architectures," because of its size but also because of its dynamicity. Secondly, the term is also used, perhaps less frequently, to refer to the set of techniques that are being developed to deal with such volumes of data. For example, Davenport et al. (2012) report that the term is used to refer to "smarter, more insightful analysis" of large volumes of data, while Oracle defines it as "techniques and technologies that enable enterprises to effectively and economically analyze all of their data" (Oracle 2012). Indeed, big data in itself is of no great value unless we find means of managing and analyzing this less conventional data, which is not necessarily formatted according to the usual rows and columns of traditional databases. The question raised by big data is therefore how to extract information and knowledge from these raw data streams, since traditional approaches are not suitable for such amount and heterogeneity of data coming from various sources (Birkin 2012; Davenport et al. 2012).

In parallel with the aforementioned technological advances for data production and collection, storage technologies have also been significantly improved, making storage relatively cheap. In 2009, writing on the *Pathologies of Big Data*, Jacobs was already saying that "transaction processing and data storage are largely solved problems" (Jacobs 2009). Rather, major difficulties arise when it comes to extracting information and learning something from massive data sets. For example, Vatsavai et al. (2012) highlight the limitations of traditional spatial data mining algorithms such as mixture models for spatial clustering or the Markov random field classifiers for land cover analysis when confronted with massive data sets.

The range of techniques and technologies dedicated to big data is wide. Existing analytic techniques for extracting knowledge from the data are being improved to be able to deal with massive data sets. These techniques include SQL queries, data mining, statistical analysis, clustering, natural language processing, text analytics, and artificial intelligence, to name a few (Russom 2011). These analytic techniques can be deployed to improve performance of knowledge extraction algorithms such as social media analysis (Mathioudakis et al. 2010), change detection algorithms for high-resolution images (Pacifici and Del Frate 2010), and complex object recognition (Vatsavai et al. 2012), for example. The ability to deal with massive data sets is supported by underlying technologies such as Google's MapReduce big data processing framework and its open-source implementation Hadoop, which is now considered by some as the de facto standard in industry and academia (Dittrich and Quiané-Ruiz 2012). MapReduce is a programming model that supports the development of scalable parallel applications for big data (Dean and Ghemawat 2008). It is based on a distributed file system where data, represented as (key, value) pairs, are initially partitioned in several machines before being processed. With MapReduce, data mining algorithms such as clustering, frequent pattern mining, classifiers, and graph analysis can be parallelized to be able to deal with massive data sets. Several ongoing researches are conducted to improve MapReduce, with for example, enhanced join algorithms (Okcan and Riedewald 2011), query optimization techniques (Herodotou and Babu 2011), and indexing techniques (Jiang et al. 2010). While still being developed, these enabling technologies will certainly play an important role in the use of VGI.

9.3 VGI AS BIG DATA

Back in 1997, Goodchild pointed out that as networks became increasingly ubiquitous, the production of GI was moving from a centralized to a distributed process. Nowadays, users can produce GI via a variety of Internet applications; as a result, a *global digital commons of geographic knowledge* is created without having to rely solely on *traditional* geospatial data production processes (Hardy 2010). In 2007, Goodchild introduced the term *volunteered geographic information* to refer to the GI generated by users through Web 2.0 era applications. Later, Ballatore and Bertolotto (2011) stated that the VGI paradigm reflects the transformation of users from *passive* geospatial information consumers to *active contributors*. However, Coleman et al. (2009) argue that the concept of *user-generated content* is not new, referring for instance to public participation GIS where users can provide input and feedback to decision makers and involved communities through web-based applications. The novelty, they claim, lies in part in the community-based aspect of the users' contribution to this digital commons of geographic knowledge (Coleman et al. 2009). VGI is often created out of the collaborative involvement of large communities of users in a common project, for example, OpenStreetMap (OSM) or Wikimapia, where individuals can produce GI that emanates from their own local knowledge of a geographic reality or edit information provided by other individuals. For example, in OSM, users can describe map features—such as roads, water bodies, and points of interest—using *tags*, providing information at a level that often goes beyond the level of detail that can be provided by traditional geospatial data producers (Goetz et al. 2012). As a result, and with the ever-increasing number of crowdsourcing applications, the volume of VGI is becoming huge, with no doubt that VGI is now an important component of big data.

Among the advantages associated with VGI, researchers highlight its use to enrich, update, or complete existing geospatial data sets (Goetz and Zipf 2011; Goodchild 2007; Gupta 2007; Tulloch 2008). This advantage is especially put forward in the context where traditional geospatial data producers, which are usually governments, may lack the capacity to generate data sets with comprehensive spatial and temporal coverage and level of detail (Gupta 2007; Tulloch 2008) such as those needed for routing services to be efficient. Furthermore, it was highlighted that VGI can be provided and disseminated in a timely, near real-time fashion, which is highly required in routing services. The advantages associated with VGI strongly suggest that this type of knowledge is highly valuable and is likely to help providing a dynamic picture of the environment (Mooney and Corcoran 2011). Nevertheless, the use of VGI for mobile applications such as routing and navigation is not yet fully achievable, as it is hampered by various obstacles related to large volumes, heterogeneity, and credibility.

9.4   TRADITIONAL ROUTING SERVICES

Vehicle routing is primarily based on digital geographic data. The off turning of selective availability of GPS by the order of the former US president Bill Clinton in the year 2000 (http://clinton3.nara.gov/WH/EOP/OSTP/html/0053_2.html) was the beginning of a new era of positioning on Earth. Since then, positioning has become more precise and mobile systems have been able to localize themselves precisely on the road. Consequently, routing is combined with navigation. The digitization of the existing streets by mainly two companies (NAVTEQ and TomTom, [formerly TeleAtlas]) provides the necessary data set for routing applications. In parallel, web services for routing were emerged. One of the most popular services is Google Maps that provides routes in many areas (http://googleblog.blogspot.de/2005/02/mapping-your-way.html). Since its launch in 2005, the service has been improved by the addition of public transport data and the integration of real-time traffic information. By developing new algorithms, especially those that incorporate hierarchical techniques, routing has become faster, particularly for long route calculations (Delling et al. 2009). Hierarchical techniques exploit the hierarchical structure of the road network to reduce the time required to compute queries (Bauer et al. 2008). Nowadays, vehicle routing can provide routes based on such criteria as distance, road type, and traffic. In terms of standards for interoperability, interfaces for location-based services (LBSs), and in particular for routing, were developed (Mabrouk et al. 2005).

With the arrival of VGI and, notably, of the OSM project, a second generation of route planning services is starting to emerge. VGI applications make routable data more easily available for free (Neis and Zipf 2008) and reduce information gaps. For example, the OpenRouteService.org uses OSM as a source of data (Figure 9.1). Users can add points of interest (POIs) as identified in OSM or search for POIs by name. Using other data sources (besides OSM) to search for POIs would enable users to

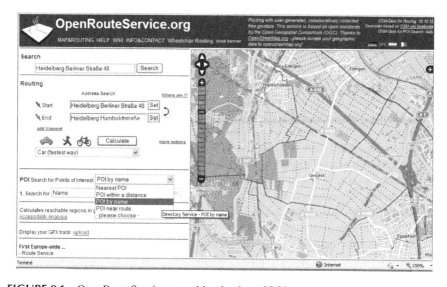

FIGURE 9.1   OpenRouteService.org with selection of POIs.

choose among a wider selection of POIs; however, extensive integration work would be required to merge the different heterogeneous data sets.

In addition, special routing services such as wheelchair routing, bike routing, and agricultural routing, to name only a few, are being designed. To provide the appropriate data required for the aforementioned types of routing services, new sources of data have been considered, for example, crowdsourced and sensor data. For instance, they help to consider weather conditions such as rain, ice, or snow, which can be measured by sensors and, therefore, the traffic conditions can be up-to-the-minute predicted. Undoubtedly, with the help of sensor data or crowdsourced information, it will be possible to consider traffic load, weather conditions, and physical road conditions to support near real-time route planning. For example, traffic jams caused by slippery roads can be avoided. But for realizing these visions, a closer look at the data is necessary.

9.5 ROUTING SERVICES USING BIG VGI/CROWDSOURCED DATA

Within a world where sensors are pervasive, we are now able to collect huge amounts of data. From weather stations, traffic information gathered by floating car data (FCD), data from vehicle telematics systems, mobile phone data, or sensor data of smart phones, the possibility for improving routing services through the use of such data is huge. Other data sources that can be of interest include social media like Twitter, Facebook, or Flickr. All of these data sets are voluminous (with size being dependent on the region). Consequently, these data sets fall under the big data paradigm because of their volume, dynamicity, and heterogeneity. As VGI services offer new types of data, for example, photos, POIs, and events, VGI has the potential to substantially improve the performance and outcomes of the existing routing services. Some examples will be explained in the following section.

9.5.1 ROUTING WITH LANDMARKS EXTRACTED FROM BIG VGI/CROWDSOURCED DATA

Landmarks have been defined as any feature that can *potentially serve as a point of reference* (Lynch 1960). They play an important role when looking for the best route, and therefore, they help to increase the relevance and personalization of routing services (Duckham et al. 2010). Example of landmarks, identified in Zhu's classification of landmarks (2012), include social landmarks, which are places where people socialize and interact, such as parks and public places; economic landmarks, such as markets; and historical landmarks, such as monuments and historic buildings. It has been shown that such landmarks are important in forming mental representations of the environment (Couclelis et al. 1987).

The integration of landmarks in navigation and routing services involves two steps. The first step is the identification of features that could be considered as landmarks. The second step is the classification of landmarks, according to different parameters that indicate relevance, and the selection of the appropriate landmarks to be included in the routing preferences. There are a number of research projects focused on extracting and ranking of landmarks for routing purposes

(Caduff and Timpf 2005; Duckham et al. 2010; Klippel and Winter 2005; Raubal and Winter 2002; Winter et al. 2008). Traditionally, landmarks are extracted from topological databases. However, other data sources have been used as well, such as web documents (Tezuka and Tanaka 2005).

Besides traditional spatial databases, VGI and crowdsourcing can be increasingly regarded as relevant sources for landmarks. Because VGI and crowdsourcing sources, in some regions, provide a more comprehensive spatial and temporal coverage and level of detail than sources provided by the governmental/commercial mapping agencies, they can complement these official sources where local features are identified from the point of view of users and local population. As an example, the Flickr photo portal is an interesting potential source for identifying landmarks. In Flickr, users can post photos, associate tags with these photos, and create groups of photos around particular themes. Photos, tags, and themes can be exploited to extract landmarks. The problem is that due to the very high volume of pictures and tags, as well as heterogeneity of tags, searching for landmarks in Flickr will result in a large number of irrelevant results. An appropriate landmark identification algorithm would need to cross-reference different information in Flickr (e.g., content-based image analysis, text, tags, social groups) in order to identify a relevant subset of landmarks. Similar problems arise when considering other potential sources of landmarks, such as OSM where objects are identified with (key, value) pairs that can use heterogeneous terms. Nevertheless, these sources have a huge potential to make more personalized routes where a large variety of landmarks can be selected according to users' profiles and interests.

9.5.2 GPS Traces

Maps used for routing services are generally created from geographical databases. Such maps are not frequently updated because of the cost associated with these updates. Recently, interest in building road maps automatically from GPS traces as a complement to geographical databases has increased (Liu et al. 2012).

GPS traces are sequences of GPS coordinates generated by a moving device (e.g., instrumented probe vehicle). GPS traces can be obtained from cell phones, in-vehicle navigation devices, or GPS-enabled devices. Currently, most active road users, such as taxis, public transports, freight services, and agricultural vehicles, use navigation systems. Some of these vehicles are tracked by proprietary telematics systems (Lauer and Zipf 2013). These telematics systems can generate massive volumes of spatiotemporal data, from which relevant information that can improve routing services can be extracted. GPS traces allow extraction of vehicle trajectories and measurement of vehicle speed, idling, and traffic congestion. Such derived data can allow generating temporally detailed road maps, where the current speed of vehicles on road segments can be known in near real time to prevent other vehicles approaching congested areas. Using data on vehicle speed derived from GPS traces, we can also estimate fuel efficiency and greenhouse gas emissions. This would enable to make personalized route recommendations where greenhouse gas emissions and fuel consumption are minimized. However, deriving such data involves accessing and analyzing massive temporal GPS data.

9.5.3 Social Media Reports

Conventional traffic data collection methods are based on roadside inductive loop detectors, which are costly to deploy and maintain. As an alternative, social media can be leveraged to gather information on road conditions. A number of applications have been developed that allow sharing road network information. A popular example is Waze, which is a community-based traffic and navigation application (https://www.waze.com/). The application can deliver alerts when road incidents such as accidents, road hazards, traffic jams, and meteorological events occur and are detected and shared by contributors. Waze integrates a traffic flow model derived from phone data in its routing engine to refine route calculation by adjusting edge weights. Another example is Traffic Pulse, a participatory mobile sensor web platform that relies on voluntary use of smart phones to report on road conditions (Li et al. 2012). This mobile application allows to querying and visualizing the city's mobility information in real time. Other types of social platforms dedicated to more specific aspects of road networks are available, such as platforms to report on potholes (e.g., Pothole Info). Reporting of road incidents by several existing social media and web platforms presents a huge potential to keep drivers and travelers informed of the best route at any given time.

9.6 CHALLENGES FOR EXPLOITING BIG VGI TO IMPROVE ROUTING SERVICES

9.6.1 Limitations of VGI and Crowdsourced Data

Credibility, reliability, and quality are among the main issues raised regarding VGI and crowdsourcing (Bishr and Mantelas 2008; Elwood 2008; Flanagin and Metzger 2008; Gouveia and Fonseca 2008; Jokar Arsanjani et al. 2013; Sieber 2007). VGI can be perceived as lacking credibility and reliability because it is produced by non-experts in a context that highly differs from the *structured institution-initiated and expert-driven contexts* (Elwood 2008). For example, while expert geospatial data producers are expected to generate data with a certain level of precision, users of VGI applications are not formally required to do so. As explained by De Longueville et al. (2010), users of VGI applications may have a vague memory of the geographic phenomenon they report on, or they could perceive only a certain portion of it. Another concern related to the quality of VGI is the fact that the profile and motivation of contributors are often unknown. As mentioned by De Longueville et al. (2010), the socioeconomic, sociological, and cultural aspects that characterize users can have an impact on VGI generation, by making it less reliable. Being aware of the relevant characteristics of the contributors can help properly interpret VGI and assess its quality and fitness for use, since evaluating the quality of geospatial data sets (e.g., VGI) in spatial data infrastructure is a challenging research problem (Mobasheri 2013).

Besides the issue of VGI quality, the GI community still faces a number of obstacles regarding how VGI can be interpreted. VGI is often produced and stored using natural language, rather than agreed-upon terminologies and formal language usually employed in existing standardized geospatial database systems (Elwood 2008).

According to Scheider et al. (2011), the terms used by contributors to describe geographic phenomena lack *unambiguous interpretation in terms of reproducible observations.* Nevertheless, some VGI applications, such as OSM, require contributors to employ a predetermined terminology (codified in an ontology). However, Scheider et al. (2011) indicate that it is difficult to reach a consensus regarding the terminology to use, while Mooney and Corcoran (2011) state that there is a lack of a mechanism for checking adherence to the agreed-upon ontology. As a result, the heterogeneity affecting VGI is likely to be more severe than the heterogeneity affecting traditional geospatial data (Grossner and Glennon 2007; Hyon 2007).

9.6.2 Impact on the Development of Routing and Navigation Services

The aforementioned limitations associated with massive VGI and crowdsourced data sets have an impact on our ability to exploit them to improve routing and navigation services.

9.6.2.1 Interoperability

In order to exploit VGI properly, various existing data sources must be conflated to routing and navigation services. However, this is not always possible due to interoperability issues. There exist a number of researches that have used VGI or crowdsourced data from a specific source for routing purposes; for example, Abassi et al. (2009) use Flickr tags and groups to identify landmarks. However, in order to improve routing services further, data fusion and data integration techniques must be applied to fuse diverse sourced data. For example, different characteristics of a landmark could be retrieved from different sources. This is currently hampered by heterogeneity of the formats, protocols, and semantics of these sources. A first avenue is therefore to use, adapt, or develop standards to support interoperability. Currently, there exist no standards to describe VGI or crowdsourcing applications. However, some research works are being conducted that suggest how to use existing standards to facilitate access to VGI and crowdsourced data. For instance, De Longueville et al. (2010) proposed a framework where OGC's Sensor Web Enablement (SWE) standards, for example, sensor observation service (SOS), would support access to VGI sources, arguing that VGI can be described just as sensor observations. As another example, Auer and Zipf (2009) explore and demonstrate how OGC open standards can be used for user-generated content and, in particular, how the OpenRouteService uses data from OSM and WFS, WPS, WMS, etc.

9.6.2.2 Finding the Right Data

Although having access to a large variety of sources creates a huge potential, it does not guarantee that the most appropriate data required for a given situation should be found easily. Certainly, VGI data contain redundant and noisy data as well, which demand for filtering and abstracting data. This is more necessary when it comes to providing more personalized routing services, where only the geographical objects of interest would be displayed on the map. Features that should be displayed on the routing and navigation map should be selected according to users' context (his/her location, purpose of travel, mode of travel, POIs, etc.). This is not only because the map should be personalized, however. Routing and navigation services are no longer limited to the 2D world but tend to increasingly include 3D representations of the environment. But experimental research reveals significant issues in using the third dimension to create effective data visualizations, thus the need to select only the elements that should be displayed, especially for mobile devices (Shepherd 2008). The need of displaying only a subset of relevant elements puts forward the need for modelling users' context appropriately for mobile applications and gathering information on users' context. The information on users' context can also be obtained through the processing of VGI and crowdsourced data sets (e.g., logs, location, historic trajectories), which involves the use of efficient data mining algorithms of various types (text mining, clustering, etc.). Information on users' context gathered from social media could also help gain insight into the drivers' behavior, habits, and desires to recommend more suited routes.

Retrieving the relevant data can be achieved in different ways, for example, through pull-based or event-detection systems. Event detection systems should especially be considered for integrating recent changes in the road network that may affect the traffic, such as construction, meteorological events, or road accidents (Bakillah et al. 2013). This requires the coupling of various sources such as social media reports discussed earlier and the routing service. Change detection algorithms can also be leveraged too to identify changes in the road network. However, they must be able to deal with high-spatial and high temporal resolution images, which introduce important constraints on computational resources. Existing spatiotemporal data mining algorithms (such as the spatial autoregressive model, Markov Random Field classifiers, Gaussian process learning and mixture models) must be improved to become more scalable and to be able to support change detection from high spatiotemporal resolution images (Vatsavai et al. 2012).

9.6.2.3　Analyzing and Interpreting Data

Analyzing raw data from such variety of sources poses challenges due to high volume and heterogeneity of VGI and crowdsourced data. Heterogeneity, at the semantic level, means that different terminologies and representations are used to describe similar features of the real world. Because crowdsourcing and VGI applications do not necessarily adhere to standard terminologies, the heterogeneity problems are huge and far from being resolved. Heterogeneities hamper the ability to analyze and fuse heterogeneous data. Ontologies are used in some official spatial data sets to facilitate semantic interoperability; however, even for official data with controlled terminology, recent research in this domain demonstrates that it is difficult to establish consensus on a single ontology and difficult as well to bridge between the different ontologies used to describe different data sets. One solution that has been explored to deal with semantics of crowdsourced data is the use of folksonomies. The term *folksonomy* is a combination of the words *folk* and *taxonomy* (Wal 2007). It is a lightweight system to share resources and the user-created ad hoc labels associated with those resources. In a folksonomy, users are free to attach any term, called a tag in folksonomy nomenclature, to the resources in question. The control and design of a taxonomy are replaced in a folksonomy with the concepts of self-organization and emergence. There are several web-based applications that demonstrate the usefulness of tags and folksonomies to gather user-generated information. These applications include Delicious for URL bookmarks, Flickr for pictures, and CiteULike for academic publications, which all use folksonomies. By enabling knowledge discovery and information filtering, folksonomies could be used to tackle the heterogeneity challenge of handling the large amount of user-generated data. However, folksonomies alone would only be one tool but could not be sufficient to resolve the heterogeneity problem, because folksonomies are heterogeneous themselves. More research is necessary to determine how folksonomies could be useful to support the fusion of different VGI sets.

Data obtained from VGI and crowdsourced applications also need to be analyzed for credibility. It is believed that such sources are not always credible because contributors are not professional and their motivation is unknown. Therefore, there is a strong need to develop methods to assess the credibility of VGI. However, it is not very likely that the motivation of individual users can be traced, since contributors are mostly anonymous. Therefore, if we cannot identify the contributors' motivations, another avenue to assess the credibility of the contributed data is corroboration, that is, verifying if such information has also been provided by another source. This is linked to the aforementioned capacity to interoperate different sources with heterogeneous formats and semantics but also to the capacity to improve the efficiency of existing knowledge extraction and data mining algorithms with big data techniques to support data exploitation. Enabling technologies such as parallel computing, cloud computing, and virtual data marts is to be explored for this purpose.

The processing and analysis of VGI and crowdsourced data are also constrained by problems regarding the privacy of the contributors. For example, GPS trace data could allow tracking individual's trajectory and traveling patterns, which is a potential threat to privacy and safety of the individual. To address this, one solution is to aggregate GPS trace data before making it available to applications.

9.6.3 Applicability of Big Data Solutions to Big VGI

The solutions that have been proposed to deal with big data have the potential to help enhance the interoperability, discoverability, analysis, and interpretation of large VGI data sets. However, solutions for big data are not sufficient alone, because, as explained in this chapter, VGI displays unique characteristics that differentiate it from other more conventional types of data. Table 9.1 examines some of the solutions that were put forward to deal with big data and explain their limitations or how they should be complemented with other techniques to be suitable for VGI.

Standardization (of information models and services) is one main solution to be able to deal with large amounts of data coming from different sources and ensure interoperability. One of the main drawbacks with respect to VGI is that it is difficult to impose standards to the VGI community. On the other hand, however, open standards being developed, for example, the OGC Web Processing Service, are more and more used by VGI applications because the *open* nature of these standards matches with the ideological objective of VGI, which is to make data *open* to anyone.

With respect to discovery and analysis capacity, the parallelization techniques such as MapReduce processing framework offer interesting potential to increase processing

TABLE 9.1

Analyzing Big Data Solutions against VGI Requirements

Big Data Solutions for Improving Interoperability, Discovery, Analysis, and Interpretation of Big VGI		
Interoperability	Discovery and Analysis	Interpretation
Standardization of information models and service interfaces: more difficult to impose on VGI than on traditional data	*Parallelization and cloud computing*: potential to improve processing capacity. VGI is noisier than traditional data, so improving processing capacity does not automatically ensure better discovery and analysis.	Linked Data (RDF graphs) Need automated or semiautomated techniques to generate semantically meaningful links

capacity for performing these tasks. However, VGI data sets are intrinsically noisier than conventional data, so intensive data management quality must still be developed for VGI in particular. A similar analysis can be conducted for cloud computing, which allows to deal with volumes of data but does not address the issue of quality.

Appropriate interpretation of data is one of the main keys to help discovery and meaningful analysis of VGI data sets. Appropriate interpretation of data requires explicit semantics associated with data. One of the solutions for giving meaning to large data sets is Linked Data. Linked Data is a web of data coming from different sources, linked through Resource Description Framework (RDF) predicates (Bizer et al. 2009). For example, in Linked Data, two entities (e.g., Department of Geomatics Engineering and University of Calgary) can be identified by their unique resource identifiers (URIs) and linked through the predicate *within*. As Linked Data contains huge amount of data sets semantically linked to other data sets, it constitutes a rich source for supporting the interpretation of data. However, to be applicable to VGI data sets, there is a need for automated or semiautomated techniques to generate semantically meaningful links between entities of VGI data sets and entities in Linked Data. In this case again, the issues of noisiness and heterogeneity are the main obstacles to establishing appropriate links.

9.7 SUMMARY

Routing and navigation services are moving from a relatively static, fit-for-all perspective to a highly personalized, near real-time routing and navigation experience. The objective of routing services is no longer just to recommend *the shortest route* but rather to recommend the route that takes into account a large variety of information and preferences. These preferences include POIs, the very current road conditions, and the optimization of various parameters such as energy consumption. In addition, routing and navigation services are no longer exclusively considered as personal services for individual consumers. They also present a collective interest. For example, while routing and navigation services can encourage drivers to choose routes that reduce energy consumption, they can also be considered as part of an integrated strategy by authorities and civil society to improve energy efficiency and to reduce atmospheric pollution.

This vision is partially driven by the emergence of VGI and crowdsourced applications, as well as by the increasing number of examples of successful use of these new types of data. However, as part of the big data paradigm, VGI and crowdsourced data are affected by characteristics that limit their use in routing and navigation services. These limitations, which were reviewed in this chapter, have an impact on the interoperability of such data, the ability to identify, filter, and retrieve relevant data for routing and navigation, and the ability to extract information from these data and to analyze and interpret them. Further research will be necessary to enable the use of standards for VGI and crowdsourcing applications that will facilitate interoperability, to extract from VGI and crowdsourced data the information on users' context and their environment by exploiting big data techniques and technologies, to enable the production of temporally detailed roadmaps with current conditions and speed on road segments, to improve semantic descriptions and processing of VGI and crowdsourced data, and to protect privacy of contributors.

REFERENCES

Abbasi, R., Chernov, S., Nejdl, W., Paiu, R., and Staab, S. 2009. Exploiting Flickr tags and groups for finding landmark photos. In: *31th European Conference on IR Research*, M. Boughanem, C. Berrut, J. Mothe, C. Soule-Dupuy (eds.), *Advances in Information Retrieval*, LNCS, vol. 5478. Springer-Verlag, Berlin, Germany, pp. 654–661.

Auer, M. and Zipf, A. 2009. How do free and open geodata and open standards fit together? From scepticism versus high potential to real applications. *The First Open Source GIS UK Conference.* University of Nottingham, Nottingham, U.K.

Bakillah, M., Zipf, A., Liang, S.H.L., and Loos, L. 2013. Publish/subscribe system based on event calculus to support real-time multi-agent evacuation simulation. In D. Vandenbroucke, B. Bucher, and J. Crompvoets (eds.), *AGILE 2013*, LNGC, Springer-Verlag, Berlin, Germany. Leuven, Belgium, pp. 337–352.

Ballatore, A. and Bertolotto, M. 2011. Semantically enriching VGI in support of implicit feedback analysis. In: K. Tanaka, P. Fröhlich, and K.-S. Kim (eds.), *Proceedings of W2GIS 2011*, LNCS, vol. 6574. Springer-Verlag, Berlin, Germany, pp. 78–93.

Bauer, R., Delling, D., Sanders, P., Schieferdecker, D., Schultes, D., and Wagner, D. 2008. Combining hierarchical and goal-directed speed-up techniques for Dijkstra's algorithm. In: C.C. McGeoch (ed.), *WEA 2008*, LNCS, vol. 5038. Springer, Heidelberg, Germany, pp. 303–318.

Birkin, M. 2012. Big data challenges for geoinformatics. *Geoinformatics and Geostatistics: An Overview*, 1(1), 1–2.

Bishr, M. and Mantelas, L. 2008. A trust and reputation model for filtering and classifying knowledge about urban growth. *GeoJournal*, 72(3–4), 229–237.

Bizer, C., Heath, T., and Berners-Lee, T. 2009. Linked data—The story so far. *International Journal on Semantic Web and Information Systems*, 5(3), 1–22.

Caduff, D. and Timpf, S. 2005. The landmark spider: Representing landmark knowledge for wayfinding tasks. In: T. Barkowsky, C. Freksa, M. Hegarty, and R. Lowe (eds.), *Reasoning with Mental and External Diagrams: Computational Modeling and Spatial Assistance—Papers from the 2005 AAAI Spring Symposium*, Stanford, CA, March 21–23, 2005. AAAI Press, Menlo Park, CA, pp. 30–35.

Coleman, D., Georgiadou, Y., and Labonté, J. 2009. Volunteered geographic information: The nature and motivation of producers. *International Journal of Spatial Data Infrastructures Research*, 4, 332–358. Special Issue GSDI-11.

Couclelis, H., Golledge, R.G., Gale, N., and Tobler, W. 1987. Exploring the anchor-point hypothesis of spatial cognition. *Journal of Environmental Psychology*, 7, 99–122.

Davenport, T.H., Barth, P., and Bean, R. 2012. How "big data" is different. *MIT Sloan Management Review*, 54(1), 22–24.

De Longueville, B., Ostlander, N., and Keskitalo, C. 2010. Addressing vagueness in volunteered geographic information (VGI)—A case study. *International Journal of Spatial Data Infrastructures Research*. Vol. 5, Special Issue GSDI-11.

Dean, J. and Ghemawat, S. 2008. MapReduce: Simplified data processing on large clusters. *Communications of the ACM*, 51(1), 107–113.

Delling, D., Sanders, P., Schultes, D., and Wagner, D. 2009. Engineering route planning algorithms. In: J. Lerner, D. Wagner, and K.A. Zweig (eds.), *Algorithmics of Large and Complex Networks*, LNCS, vol. 5515. Springer, Berlin, Germany, pp. 117–139. doi:10.1007/978-3-540-72845-0_2.

Dittrich, J. and Quiané-Ruiz, J.-A. 2012. Efficient big data processing in Hadoop MapReduce. *Proceedings of the VLDB Endowment*, 5(12), 2014–2015.

Duckham, M., Winter, S., and Robinson, M. 2010. Including landmarks in routing instructions. *Journal of Location-based Services*, 4(1), 28–52.

Dumbill, E. 2013. Making sense of big data. *Big Data*, 1(1), 1–2.

Elwood, S. 2008. Volunteered geographic information: Future research directions motivated by critical, participatory, and feminist GIS. *GeoJournal*, 72(3), 173–183.

Flanagin, A. and Metzger, M. 2008. The credibility of volunteered geographic information. *GeoJournal*, 72(3), 137–148.

Goetz, M., Lauer, J., and Auer, M. 2012. An algorithm-based methodology for the creation of a regularly updated global online map derived from volunteered geographic information. In: C.-P. Rückemann and B. Resch (eds.), *Proceedings of Fourth International Conference on Advanced Geographic Information Systems, Applications, and Services*, Valencia, Spain, IRIA, pp. 50–58.

Goetz, M. and Zipf, A. 2011. Towards defining a framework for the automatic derivation of 3D CityGML models from volunteered geographic information. *Joint ISPRS Workshop on 3D City Modelling & Applications and the Sixth 3D GeoInfo Conference*, Wuhan, China.

Goodchild, M.F. 1997. Towards a geography of geographic information in a digital world. *Computers, Environment and Urban Systems*, 21(6), 377–391.

Goodchild, M.F. 2007. Citizens as sensors: The world of volunteered geography. *GeoJournal*, 69(4), 211–221.

Gouveia, C. and Fonseca, A. 2008. New approaches to environmental monitoring: The use of ICT to explore volunteer geographic information. *GeoJournal*, 72(3), 185–197.

Grossner, K. and Glennon, A. 2007. Volunteered geographic information: Level III of a digital earth system. *Position Paper Presented at the Workshop on Volunteered Geographic Information*, Santa Barbara, CA, December 13–14, 2007, 2pp.

Gupta, R. 2007. Mapping the global energy system using wikis, open sources, www, and Google Earth. *Position Paper Presented at the Workshop on Volunteered Geographic Information*, Santa Barbara, CA, December 13–14, 2007, 2pp.

Hardy, D. 2010. Volunteered geographic information in Wikipedia. PhD thesis, University of California, Santa Barbara, CA, 260pp.

Herodotou, H. and Babu, S. 2011. Profiling, what-if analysis, and cost-based optimization of MapReduce programs. *PVLDB*, 4(11), 1111–1122.

Hyon, J. 2007. Position paper on "specialist meeting on volunteered geographic information." *Position Paper Presented at the Workshop on Volunteered Geographic Information*, Santa Barbara, CA, December 13–14, 2007, 2pp.

Jacobs, A. 2009. The pathologies of big data. *Communications of the ACM*, 52(8), 36–44.

Jiang, D., Chin Ooi, B., Shi, L., and Wu, S. 2010. The performance of MapReduce: An in-depth study. *Proceedings of the VLDB Endowment*, 3(1–2), 472–483.

Jokar Arsanjani, J., Barron, C., Bakillah, M., and Helbich, M. 2013. Assessing the quality of OpenStreetMap contributors together with their contributions. *Proceedings of the 16th AGILE Conference*, Leuven, Belgium.

Klippel, A. and Winter, S. 2005. Structural salience of landmarks for route directions. In: A.G. Cohn and D. Mark (eds.), *Spatial Information Theory*. Lecture Notes in Computer Science. Springer, Berlin, Germany, pp. 347–362.

Lauer, J. and Zipf, A. 2013. Geodatenerzeugung aus landwirtschaftlichen Telematikdaten—Neue Methoden zur Verbesserung nutzergenerierter Daten am Beispiel TeleAgro+. *Proceedings of the Geoinformatik*, Heidelberg, Germany.

Li, R.-Y., Liang, S.H.L., Lee, D.W., and Byon, Y.-J. 2012. TrafficPulse: A green system for transportation. *The First ACM SIGSPATIAL International Workshop on Mobile Geographic Information Systems 2012 (MobiGIS2012)*. ACM Digital Library, Redondo Beach, CA.

Liu, X., Biagioni, J., Eriksson, J., Wang, Y., Forman, G., and Zhu, Y. 2012. Mining large-scale, sparse GPS traces for map inference: Comparison of approaches. *KDD'12*, Beijing, China, August 12–16, 2012.

Lohr, S. 2012. The age of big data. *The New York Times*, February 11, 2012.

Lynch, K. 1960. *The Image of the City*. MIT Press, Cambridge, U.K.

Mabrouk, M., Bychowski, T., Williams, J., Niedzwiadek, H., Bishr, Y., Gaillet, J.-F., Crisp, N. et al. 2005. OpenGIS Location Services (OpenLS): Core Services, OpenGIS® implementation specification. OpenGeospatial consortium. Retrieved from http://portal.opengeospatial.org/files/?artifact_id = 3839&version = 1 (accessed April 2013).

Mathioudakis, M., Koudas, N., and Marbach, P. 2010. Early online identification of attention gathering items in social media. *Proceedings of the Third ACM International Conference on Web Search and Data Mining*. ACM, New York, pp. 301–310. doi: 10.1145/1718487.1718525.

Mobasheri, A. 2013. Exploring the possibility of semi-automated quality evaluation of spatial datasets in spatial data infrastructure. *Journal of ICT Research and Applications*, 7(1), 1–14.

Mooney, P. and Corcoran, P. 2011. Annotating spatial features in OpenStreetMap. *Proceedings of GISRUK 2011*, Portsmouth, U.K., April 2011.

Neis, P. and Zipf, A. 2008. Zur Kopplung von OpenSource, OpenLS und OpenStreetMaps in OpenRouteService.org. *Proceedings of Symposium für angewandte Geoinformatik*, Salzburg, Austria, 2008.

Okcan, A. and Riedewald, M. 2011. Processing theta-joins using MapReduce. *Proceedings of the 2011 ACM SIGMOD International Conference on Management of Data*, Athens, Greece, June 12–16, 2011. doi: 10.1145/1989323.1989423.

Oracle. 2012. Big data technical overview. Available at: http://www.oracle.com/in/corporate/events/bigdata-technical-overview-pune-1902240-en-in.pdf (accessed March 30, 2013).

Pacifici, F. and Del Frate, F. 2010. Automatic change detection in very high resolution images with pulse-coupled neural networks. *IEEE Geoscience and Remote Sensing Letters*, 7(1), 58–62.

Raubal, M. and Winter, S. 2002. Enriching wayfinding instructions with local landmarks. In: M. Egenhofer and D. Mark (eds.), *Geographic Information Science*. Lecture Notes in Computer Science. Springer, Berlin, Germany, pp. 243–259.

Russom, P. 2011. Big data analytics. TDWI best practices report. TDWI Research.

Scheider, S., Keßler, C., Ortmann, J., Devaraju, A., Trame, J., Kauppinen, T., and Kuhn, W. 2011. Semantic referencing of geosensor data and volunteered geographic information. In: N. Ashish and A.P. Sheth (eds.), *Geospatial Semantics and the Semantic Web*. *Semantic Web and Beyond*, vol. 12. Springer, New York, pp. 27–59.

Shepherd, I.D. 2008. Travails in the third dimension: A critical evaluation of three-dimensional geographical visualization. In: M. Dodge, M. McDerby, and M. Turner (eds.), *Geographic Visualization: Concepts, Tools and Applications*. Wiley, Chichester, U.K., pp. 199–222.

Sieber, R. 2007. Geoweb for social change. *Position Paper Presented at the Workshop on Volunteered Geographic Information*, Santa Barbara, CA, December 13–14, 2007, 3pp.

Tezuka, T. and Tanaka, K. 2005. Landmark extraction: A web mining approach. In: A.G. Cohn and D.M. Mark (eds.), *Spatial Information Theory*. Lecture Notes in Computer Science. Springer, Berlin, Germany, pp. 379–396.

Tulloch, D. 2008. Is volunteered geographic information participation? *GeoJournal* 72(3), 161–171.

Vatsavai, R.R., Chandola, V., Klasky, S., Ganguly, A., Stefanidis, A., and Shekhar, S. 2012. Spatiotemporal data mining in the era of big spatial data: Algorithms and applications. *ACM SIGSPATIAL BigSpatial'12*, Redondo Beach, CA, November 6, 2012.

Wal, T.V. 2007. Folksonomy. http://vanderwal.net/folksonomy.html (accessed April 2013).

Winter, S., Tomko, M., Elias, B., and Sester, M. 2008. Landmark hierarchies in context. *Environment and Planning B Planning and Design*, 35(3), 381–398.

Zhu, Y. 2012. Enrichment of routing map and its visualization for multimodal navigation. Doctoral thesis, Technische Universität Müchen, Munich, Germany, 154pp.

10 Efficient Frequent Sequence Mining on Taxi Trip Records Using Road Network Shortcuts

Jianting Zhang

CONTENTS

10.1 INTRODUCTION

Locating and navigation devices now are ubiquitous. Huge amounts of geo-referenced spatial location data and moving object trajectory data are being generated at ever-increasing rates. Patterns discovered from these data are valuable in understanding human mobility and facilitating traffic mitigation. The state-of-the-art techniques in managing such data are based on spatial databases and moving object databases to index and query geometrical coordinates directly. However, as most of human movements are constrained by built infrastructures and road networks, an alternative approach to matching the location and trajectory data with infrastructure data for subsequent processing (Richter et al. 2012) can potentially be more efficient from a computing perspective. For example, once a global positioning system (GPS) trajectory is transformed into a sequence of road segments, it will be possible to apply well-studied frequent sequence mining (FSM) algorithms (Agrawal and Srikant 1995) to identify popular routes for different groups of people at different scales for different purposes, for example, ridesharing, hailing taxies for riders, making more profits for drivers, and understanding functional zones in metropolitan areas to facilitate city planning and traffic mitigation.

193

While the idea is attractive, besides the challenge in accurate map matching, our preliminary results have shown that applying frequency counting-based association rule mining algorithms in a straightforward manner may incur significant computing overheads due to the inherent combinatorial complexity of the algorithms. For large metropolitan areas such as New York City (NYC) with a road network of hundreds of thousands of intersections and segments and millions of trips a day, the computing overheads quickly become intractable. By taking advantage of the fact that many human trips follow shortest path principles in general (Eisner et al. 2011), we propose to utilize street hierarchies through network contractions (Geisberger et al. 2008) to effectively reduce the lengths of road segment sequences after map matching of trip trajectories. Shortcuts derived from road networks naturally partition sequences of road segments into subsequences. These subsequences with high frequencies are likely to be the whole or part of frequent sequences. The complexity of counting the frequencies of the shortcuts, on the other hand, is linear with the number of original and shortcut road segments in the subsequences that is much more efficient than classic sequence mining algorithms. When appropriate, classic sequence mining algorithms can be applied on the frequent subsequences to discover longer frequent sequences. As numbers of segments in shortcut sequences are typically much smaller than the numbers of original sequences, the computing workloads of sequence mining algorithms on shortcut sequences are much lighter. We empirically evaluate the proposed approach on a subset of taxi trip records in NYC in 2009 (about 168 million records in total) and report our preliminary results.

The rest of this chapter is arranged as the following: Section 10.2 introduces background, motivation, and related work. Section 10.3 presents the prototype system architecture and implementation details of several components. Section 10.4 provides experiment results. Section 10.5 concludes this chapter and provides future work directions.

10.2   BACKGROUND, MOTIVATION, AND RELATED WORK

Processing GPS data for various purposes has attracted significant research interests, such as compression, segmentation, indexing, query processing, and data mining (Zheng and Zhou 2011). Most current research is focused on the geometrical aspects, such as range, nearest neighbor and similarity queries, clustering and data mining of individual, and group-based patterns, for example, convey and flock (Zheng and Zhou 2011). Many approaches are generic for many types of moving object trajectory data but can incur heavy computation overheads. Approaches in map matching of GPS points to road network segments (Brakatsoulas et al. 2005), although mostly developed for navigation purposes, are particularly relevant to our research as our approach relies on such techniques to transform point sequences into road segment identifier sequences for frequent pattern mining. Several previous research efforts have proposed to develop semantic data models to derive individual activities from GPS trajectories (Yan et al. 2011, Richter et al. 2012). While our approach can be adapted to efficiently mine frequent sequences from GPS trajectories of individuals, our focus in this study is to understand frequently used paths from large-scale taxi trip data at the city level that is much more computationally demanding, and efficiency is overwhelmingly important.

Existing work on city-level trajectory data analytics can be grouped into several categories. Research in the first category focuses on identifying popular routes from trajectories (Chen et al. 2011, Wei et al. 2012) with applications to ridesharing recommendations (He et al. 2012). The second category focuses on identifying anomalous trajectories (Zhang et al. 2011, Chen et al. 2013) with applications to taxi driving fraud detection (Ge et al. 2011). The third category is related to identifying region of interests (ROIs) and their functions (Qi et al. 2011, Uddin et al. 2011, Huang and Powell 2012, Yuan et al. 2012). The fourth category of research mainly is interested in analyzing origin–destination (O–D) flows over space and time (Zheng et al. 2011, Jiang et al. 2012, Yuan et al. 2012). Finally, research in the fifth category is interested in deriving sophisticated knowledge from taxi GPS traces for taxi drivers (Li et al. 2011, Powell et al. 2011, Yuan et al. 2011a) and general drivers (Ziebart et al. 2008, Yuan et al. 2010, 2011b). Among these studies, except for Chen et al. (2011) that adopted a network expansion approach, the majority of existing research adopted a uniform grid-based approach, and many of the grids are quite coarse. Although using coarse resolution grids significantly reduces computational complexity, it may also significantly reduce accuracy. A possible improvement is to first use fine resolution grids and then to aggregate grid cells into regions to reduce computational overheads and maintain accuracy simultaneously (Yuan et al. 2012). It is somehow surprising that, while GPS trajectories are strongly constrained by road networks, very few works incorporate road network topology explicitly. Despite significant progress in map matching (Zheng and Zhou 2011, Ali et al. 2012), the technical difficulties, computational overheads, and qualification of the matching results might be some key factors in impeding their practical applications on large-scale GPS trajectory data. Nevertheless, we believe that the synergized hardware and software advances will significantly improve map-matching accuracy. In this study, we assume GPS trajectories are matched to road segments, and we will focus on frequent trajectory pattern mining from road segment identifier sequences.

Frequent trajectory pattern mining from road segment identifier sequences is naturally abstracted as a frequent pattern mining problem (Han et al. 2007). Since the Apriori-based association rule mining approaches for FSM were developed nearly two decades ago (Agrawal and Srikant 1995), significant progress has been made in the important data mining area. New strategies such as Eclat and FP-growth, new variations such as closed and maximal itemsets and sequences, and new hardware platforms such as clusters, multi-core CPUs, and many-core graphics processing units (GPUs) have been extensively explored (Hipp et al. 2000, Han et al. 2007, Borgelt 2012). While Borgelt (2012) argues that there is little room left for speed improvements of existing efficient and optimized implementations of frequent pattern mining, we believe exploring domain knowledge to effectively reduce problem size and number of frequent patterns can be a new growing point for empowering existing data mining algorithms for large-scale datasets. Using road hierarchy derived from network contractions is our first attempt based on this idea. As our prototype system uses the Sequoia* open source FSM package, we next provide a brief introduction to FSM based on the implementation in Sequoia.

---

* http://www.borgelt.net/sequoia.html

Given a set of sequences, FSM algorithms aim at finding unique sets of ordered itemsets whose frequencies are above a predefined threshold termed as support value, which can be either an absolute value or a percentage. The task becomes simpler when only one item is allowed in the itemsets that are used as the building blocks for frequent sequences. While allowing multiple items in itemsets provides higher flexibility, using a single item per itemset may actually be more suitable for trajectory data where road segments are typically unique in simple frequent sequences. The Sequoia package by Borgelt is highly engineered from a software development perspective. It is designed to find both non-closed and closed frequent sequences with unique occurrences of items, which matches our need on mining frequent sequences of unique road segments very well. Sequoia adopts the FP-growth style strategy (Han et al. 2000) to efficiently prune the search space. Our experiments also show that Sequoia is faster than alternatives, likely due to incorporating proven-effective data structures and highly efficient memory and disk I/Os, in addition to adopting efficient mining algorithms.

Using betweenness centralities (BCs) (Brandes 2008) provides another way to understand the utilization and popularity of nodes and edges in a street network (Kazerani and Winter 2009, Leung et al. 2011). More formally, the edge-based BC $C_B(e)$ is defined as the following:

$$C_B(e) = \sum_{s \neq v \neq t \in V} \frac{\sigma_{st}(e)}{\sigma_{st}}$$

where s and t are the source and destination nodes in a path and function $\sigma(e)$ accumulates numbers of paths that pass through edge e. As σ_{st} is typically a constant for road networks, we use $\sum \sigma_{st}(v)$ directly in this study. Compared with association rule-based FSM algorithms, it is much simpler to compute edge centralities as the complexity is bounded by the total number of edges in the path set. However, it is easy to see that the BC measurement is per-segment based and does not identify frequent sequences directly. Nevertheless, when BCs for a set of road segments are visualized in a map, as neighboring segments typically have similar BCs, users might be able to identify consecutive road segment sequences with high BCs in a visual manner. As shall be clear later, our approach provides a natural trade-off between per-segment-based BC measurement and per-sequence-based association mining by using shortcuts as the basic units. The shortcuts are subsequences of original sequences but require only linear scanning to count their frequencies just like the between centrality measurement. As the shortcuts are derived through the contraction hierarchy (CH) technique, we next provide more details on it.

CHs are a technique to speed up shortest path computation (SPC) (Geisberger et al. 2008). The key idea is to heuristically order the network nodes by some measurements of importance and contract them in this order. While network nodes are removed by contraction, shortcuts are added to the network being explored so that shortcut nodes do not need to be visited during node expansion in computing shortest paths. These shortcuts allow far fewer node visits in computing shortest paths than the classic Dijkstra's shortest path algorithm (Geisberger et al. 2008).

Furthermore, the nodes ordered based on their importance are much more cache and parallelization friendly during edge expansion. The data layout of network nodes and edges based on the order matches modern parallel hardware architectures very well. Experiments have shown that SPC using the CH technique on modern multi-core CPUs and many-core GPUs can be up to three orders of magnitude faster than serial Dijkstra's algorithm (Delling et al. 2011). While the CH technique is primarily designed for speeding up SPC, as human movements typically follow the shortest path principles (Eisner et al. 2011), the constructed node hierarchy can be used not only for routing purposes but also for reducing the lengths of trajectory sequences by incorporating shortcuts. As many FSM algorithms are combinatorial in nature and NP-hard in worst cases, reducing the numbers of data items in sequences can significantly reduce runtimes of frequent mining algorithms. Furthermore, mined frequent sequences consisting of important nodes naturally provide simplified frequent paths that are more informative, easy to visualize and interpret, and thus more preferable to end users.

To provide a better understanding of the CH technique, Figure 10.1 illustrates a road network in Texas with 62 nodes and 120 edges. The nodes are symbolized based on their importance after applying the technique. The more important the nodes are, the larger circles are used for symbolization. Clearly, node 46 in the center of the network is the most important node, while many nodes that are close to the boundary are least important. This is understandable, as shortest paths will likely pass through node 46 and are least likely to be shortcut. When computing the shortest path between node 34 and node 60, the full path sequence would be 34,36,57,61,22,23,53,50,11,60 whose length is 9 hops. Using the derived CH, there will be a shortcut between nodes 57 and 22, and node 61 is shortcut as it has a lower

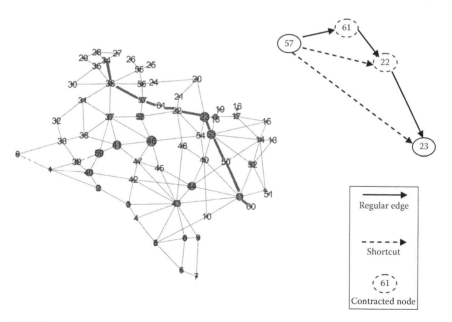

FIGURE 10.1   CHs using a Texas road network.

importance than the two nodes. Similarly, a shortcut between node 11 and node 53 is added as node 50 has a lower importance. It is also interesting to observe that there is also a shortcut between node 57 and node 23 where node 22 is shortcut in this case. The node sequence using the shortcuts turns out to be 34,36,57,23,53,11,60 whose length is 6 hops. While only 3 out of the 9 edges are reduced due to using shortcuts in this small network with a maximum level of 10, as the number of levels generally increases logarithmically with the number of nodes in a network (Geisberger et al. 2008), the reduction can be significant for long path sequences in large networks, for example, the North America road network with tens of millions of nodes and edges.

10.3 PROTOTYPE SYSTEM ARCHITECTURE

Although there are some fundamental research opportunities in tightly integrating frequent pattern mining and network contraction algorithms, as a first step, we integrate several existing packages to realize our approach in this study. Our prototype system architecture is shown in Figure 10.2 where the shaded boxes indicate input data. The U²SOD-DB data management infrastructure (Zhang et al. 2012) is used in the prototype. We currently use the CH module in the Open Source Routing Machine project* for constructing road network hierarchy and the Sequoia open source FSM package for mining frequent sequence. The trajectory to sequence (T2S) converter module is responsible for mapping GPS trajectories to sequences of road segments. The sequence contraction (SC) module is designed to contract the original sequences. These two modules are currently placeholders (indicated by the dashed frames) as the data in our experiment have only O–D locations, while the intermediate GPS readings are not available. We have developed the third module, that is, SPC, to compute shortest paths between O–D location pairs. The computed shortest paths are subsequently used as the input data for FSM. Both the original shortest path sequences and shortcut shortest path sequences are output, and they can be

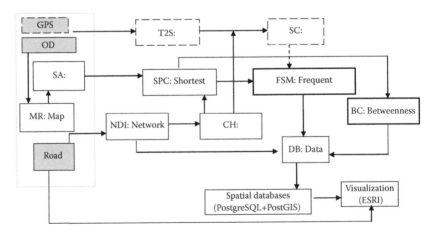

FIGURE 10.2 Prototype system architecture and components.

---

* http://project-osrm.org/

fed into the FSM module (using Sequoia) to mine frequent sequences. We have also implemented the edge-based BC computing module. By taking into account the original shortest path sequences, the BC module computes the BC measurement for each original road segments. Similarly, by taking into account the shortcut shortest path sequences, the BC module computes the BC measurement for each shortcut. The FSM and the BC modules in Figure 10.2 are highlighted in bold to indicate that they take both original and shortcut sequences.

Several auxiliary modules for preprocessing and post-processing are also developed. In particular, the network data ingesting (NDI) module is developed to ingest network data that may be provided in different formats. The module currently ingests a road network dataset from the NYC Department of City Planning (DCP), which is referred to as DCPLION and is used in our experiments. The module also supports NAVTEQ street network in the NYC area. In addition to the major streets in the five boroughs of NYC, the NAVTEQ data also cover several neighboring counties that are good for studying taxi trips whose origin or destination locations are outside of NYC's five boroughs. In addition to generating node and edge lists and computing different types of edge weights that are required by the CH module, the NI module also adds node and edge identifier columns so that nodes and edges in the sequence mining results can be mapped back to their geometry later for visualization purposes. Subsequently, a Data Bridger (DB) module is developed to combine the node and edge identifiers that are output by the Sequoia module and the related geometric data to generate structured query language (SQL) statements. The PostgreSQL/ PostGIS* spatial database can take the SQL statements, populate tables, and generate Environmental System Research Institute (ESRI) shapefiles (ESRI 1998) that can be visualized in many geographical information system (GIS) software, including ESRI's ArcGIS.[†] The DB module will be replaced by an integrated module in the future to visualize FSM results without relying on third-party programs.

10.4 EXPERIMENTS AND RESULTS

The more than 13,000 GPS-equipped medallion taxicabs in NYC generate nearly half a million taxi trips per day and more than 168 million trips per year serving 300 million passengers in 2009. The number of yearly taxi riders is about 1/5 of that of subway riders and 1/3 of that of bus riders in NYC, according to MTA ridership statistics (Metropolitan Transportation Authority 2013). Taxi trips play important roles in everyday lives of NYC residents (or any major city worldwide). Understanding the trip purposes is instrumental in transportation modeling and planning. In our previous work, we were able to compute the shortest paths between all the 168 million O–D pairs in within 2 h using a single Intel Xeon E5405 CPU (2.0 GHz) by utilizing the CH technique. The performance is orders of magnitude faster than the state of the art (Zhang 2012). However, the computed shortest path sequences are too voluminous to manipulate (with respect to streaming disk-resident files among different modules) and too cumbersome for visualization (data and

---

* http://postgis.net/
[†] http://www.esri.com/software/arcgis

patterns are cluttered when displays have limited screen resolutions). As a result, we selected a subset of 2009 data for a case study.

We are particularly interested in the taxi trips between Manhattan and the LaGuardia airport (LGA) located in Northern Queens, another borough of NYC. As one of the three major airports in the NYC metro area, the airport accommodated 24.1 and 25.7 million passengers in 2011 and 2012 (Port Authority of New York & New Jersey 2013), respectively. We extracted all the taxi trips from all the 13 community districts in Manhattan to community district 403 in Queens (where LGA is located) in January 2009, and the number of trips is 17,558. As discussed in Section 10.3, we computed the shortest paths between the pickup and drop-off locations of the trips and output the segment identifiers in the computed shortest paths for association rule-based FSM. We also applied the BC module to compute the edge-based BCs for all the edges using the original shortest path sequences. To test the effectiveness of our proposed approach, we output the shortcut sequences and apply the BC module to compute the edge-based BCs for all the shortcuts. These shortcuts are unpacked in the DB module for subsequent visualization purposes. We next present the experiments and results for the three modules, and compare them where appropriate.

10.4.1 RESULTS OF BC ON ORIGINAL SEQUENCES

Figure 10.3 visualizes the distributions of edge-based BCs with non-involving edges (BC = 0) grayed out. The figure clearly shows that the road segments of the four bridges and tunnels play important roles in the traffic between Manhattan and the airport area

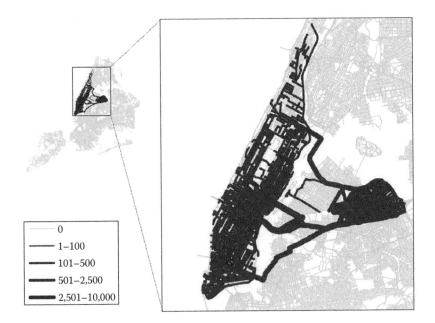

FIGURE 10.3 Visualization of edge-based BC of original shortest paths.

TABLE 10.1

Results of FSM Using Sequoia with Different Support Values

Support	Runtime (s)	# of Candidate Frequent Sequences	# of Frequent Sequences	# of Average Segments
50	0.12	3,624	6	20
25	4.42	5,470	36	38
10	37.64	10,077	275	41
5	129.64	12,771	1,018	57

as indicated by their high BCs. There are 2,620,636 road segments in the 17,558 shortest paths (~150 segments per path on average) that only take a fraction of a second to compute BCs. While the approach is efficient and provides valuable information for visualization purposes, it cannot be used to identify frequent sequences directly.

10.4.2 RESULTS OF ASSOCIATION RULE MINING ON ORIGINAL SEQUENCES

Table 10.1 lists the runtimes and several other measurements output by the Sequoia package using support values ranging from 50 to 5. Here, *candidate frequent sequences* refer to sequences consist of no road segments whose frequencies are below the respective support value and thus cannot be pruned by single item frequency counting. The candidate frequent sequences need to be fed into the FP-growth algorithm implemented in Sequoia for sequence expanding and pruning. As seen in Table 10.1, the runtimes grow superlinearly as the support values decrease. The computing is orders of magnitude slower than computing BCs. The numbers of frequent sequences also increase significantly as the support values decrease and many of them overlap. The numbers of average segments in the identified frequent sequences, however, do not increase as significantly as the runtimes and the numbers of frequent sequences. The results indicate that, as the support values become lower, both the numbers of frequent sequences and the average numbers of segments of the frequent sequences increase. For this particular dataset, the total number of frequent segments grows 3–4 orders (6*20 vs. 1018*57), although the support value only decreases 10 times (50 vs. 5). In addition, although the resulting frequent sequences do provide useful information for further processing, they are not suitable for interactive visualization as there are significant degrees of overlap among the frequent road segments, especially when support values are low.

10.4.3 RESULTS OF THE PROPOSED APPROACH

Recall that our proposed approach computes the edge-based centralities of shortcuts and uses frequent shortcuts as the approximations of frequent sequences. The approach also allows to apply the classic association rule-based FSM algorithm to further identify sequences of shortcuts with lower computing overheads. Among the 17,558 shortest paths, 2,204 shortcuts are derived. There are 149,536 segments of either original road segments or shortcuts among the 17,558 shortest paths. The average number of segments per shortest path is thus less than nine. Compared with the original shortest path sequences with about 150 segments per path on average (cf. Section 10.4.1), the average path length is reduced by almost 16 times. The reduction is considerably significant with respect to both storage and frequent pattern mining. By significantly reducing the number of segments in frequent sequences and overlap among frequent sequences, as shown in Figure 10.4, it is easier to visualize and interpret the identified frequent sequences in a GIS environment (ESRI ArcMap, in particular).

To better understand the derived shortcuts, the distributions as well as major statistics are shown in Figure 10.5. We can see that, while the average number of road

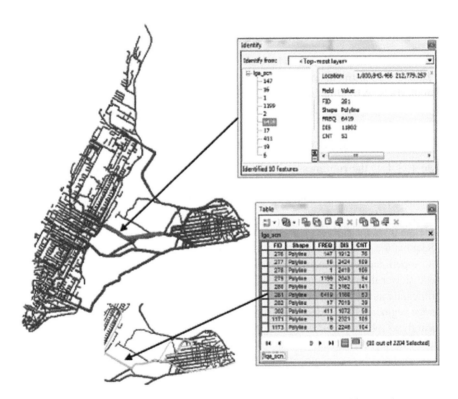

FIGURE 10.4   Examples of visualizing identified frequent sequences (shortcuts).

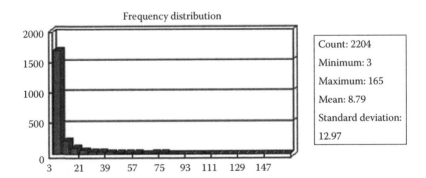

FIGURE 10.5   Frequency distribution against number of road segments in shortcuts.

segments per shortcut is close to nine, there are a few shortcuts that have large numbers of road segments. This may be partially due to the unique road network topology in NYC where the connections between Manhattan and Queens (and also the rest of the three boroughs) are mostly through a limited number of bridges and tunnels. Expressways (e.g., Interstate 495) that connect the bridges/tunnels and the airport area are likely to be major parts of the shortest paths for the taxi trips between the origins and destinations. While it is still possible to discover frequent sequences that consist of more than one shortcut, it is unlikely that all the shortcuts in such sequences have large numbers of road segments. For shortcuts with long sequences of road segments, it might be advantageous to visually explore them directly and exclude them from FSM on such shortcuts.

By applying the Sequoia module on the 17,558 shortest paths with shortcuts, our approach is able to compute frequent sequences in negligible runtime for all support values (as low as one). This is much more efficient than mining the original shortest path sequences, which took more than 2 min with the support value set to 5 as shown in Table 10.1. We note that mining the shortcut sequences will likely miss frequent sequences in several cases when compared with mining original sequences directly, such as those that cover only parts of the road segments in neighboring shortcuts. Another case might be subsequences that overlap with multiple shortcuts. When the shortcuts are not qualified as frequent sequences, the subsequences will not be identified as frequent ones within individual shortcuts. However, subsequences might be frequent across multiple shortcuts that can be identified if mining on the original sequences directly. We are working on techniques that can avoid or reduce the chances of such cases, for example, by limiting the lengths of road segments in shortcuts. We will report the efficiency and effectiveness of these techniques in our future work.

10.5 CONCLUSION AND FUTURE WORK

In this study, we have proposed a new approach to mining frequent patterns from large-scale GPS trajectory data after mapping GPS traces to road network segments. Instead of applying association rule-based FSM algorithms directly that generally have high computation overheads and are not scalable, our approach utilizes the inherent hierarchies of road networks. After contracting nodes and creating shortcuts by applying CH algorithms, the original road segment sequences are transformed into sequences of shortcuts, and the data volumes of the original sequences are significantly reduced. Edge-based BC measurements can be directly computed on shortcuts efficiently with a linear time complexity, and FSM algorithms can be applied to the shortcut sequences to identify frequent patterns with significantly reduced complexity. By using computed shortest paths as simulated GPS trajectories, our experiments on 17,558 taxi trip records have shown that computing BC measurements of shortcuts and FSM on shortcut sequences incurs negligible computational overheads. The runtimes of FSM on shortcut sequences are orders of magnitude faster than on original road segment sequences. In addition, frequent subsequences in shortcuts are more informative and interpretable based on the BCs of the shortcuts than visualizing BCs of individual road segments.

The reported work naturally leads to several future research directions. First, we would like to evaluate and validate the assumption that real-world GPS trajectories in urban areas follow the shortest path principles with different metrics, especially for long trajectories. While we currently do not have access to complete GPS traces in NYC, we plan to use the publically available T-Drive dataset* for this purpose. Second, among the 168 million taxi trips in NYC in 2009, we have used only a small fraction of the dataset, and we plan to evaluate our proposed approach on larger subsets to test its efficiency and scalability. Finally, the prototype we have developed so far is loosely coupled in nature. We plan to tightly integrate the essential components in the prototype and automate the big data analytics process. We strongly believe that by integrating road network hierarchies into frequent pattern mining algorithms, specialized and highly efficient trajectory FSM algorithms can be developed and tailored for parallel computing.

REFERENCES

Agrawal, R. and Srikant, R. 1995. Mining sequential patterns. *Proceedings of the IEEE Eleventh International Conference on Data Engineering, ICDE*, Washington, DC, pp. 3–14.

Ali, M., Krumm, J. et al. 2012. ACM SIGSPATIAL GIS Cup 2012. *Proceedings of the ACM International Conference on Advances in Geographic Information Systems, ACM-GIS*, Redondo Beach, CA, pp. 597–600.

Borgelt, C. 2012. Frequent item set mining. *Wiley Interdisciplinary Reviews: Data Mining and Knowledge Discovery*, 2(6), 437–456.

Brakatsoulas, S., Pfoser, D. et al. 2005. On map matching vehicle tracking data. *Proceedings of the 31st International Conference on Very Large Data Bases, VLDB*, Trondheim, Norway, pp. 853–864.

Brandes, U. 2008. On variants of shortest-path betweenness centrality and their generic computation. *Social Networks*, 30, 136–145.

Chen, C., Zhang, D. et al. 2013. iBOAT: Isolation-based online anomalous trajectory detection. *IEEE Transactions on Intelligent Transportation Systems*, 14(2), 806–818.

Chen, Z., Shen, H.-T., and Zhou, X. 2011. Discovering popular routes from trajectories. *Proceedings of the IEEE International Conference on Data Engineering, ICDE*, Washington, DC, pp. 900–911.

Delling, D., Goldberg, A. V. et al. 2011. PHAST: Hardware-accelerated shortest path trees. *Proceedings of the 2011 IEEE International Parallel & Distributed Processing Symposium, IPDPS*, Washington, DC, pp. 921–931.

Eisner, J., Funke, S. et al. 2011. Algorithms for matching and predicting trajectories. *Proceedings of the Workshop on Algorithm Engineering and Experiments, ALENEX*, San Francisco, CA, pp. 84–95.

ESRI. 1998. ESRI shapefile technical description. Online at http://www.esri.com/library/whitepapers/pdfs/shapefile.pdf. Last accessed October 29, 2013.

Ge, Y., Xiong, H. et al. 2011. A taxi driving fraud detection system. *Proceedings of IEEE International Conference on Data Mining, ICDM*, Washington, DC, pp. 181–190.

Geisberger, R., Sanders, P. et al. 2008. Contraction hierarchies: Faster and simpler hierarchical routing in road networks. *Proceedings of the Seventh International Conference on Experimental Algorithms, WEA*, Provincetown, MA, pp. 319–333.

---

* http://research.microsoft.com/apps/pubs/?id=152883

Han, J., Cheng, H. et al. 2007. Frequent pattern mining: Current status and future directions. *Data Mining and Knowledge Discovery*, 15(1), 55–86.

Han, J., Pei, J., and Yin, Y. 2000. Mining frequent patterns without candidate generation. *Proceedings of the 2000 ACM SIGMOD International Conference on Management of Data, SIGMOD*, Dallas, TX, pp. 1–12.

He, W., Li, D. et al. 2012. Mining regular routes from GPS data for ridesharing recommendations. *Proceedings of the ACM SIGKDD International Workshop on Urban Computing, UrbComp*, Beijing, China, pp. 79–86.

Hipp, J., Guntzer, U., and Nakhaeizadeh, G. 2000. Algorithms for association rule mining— A general survey and comparison. *SIGKDD Exploration Newsletter*, 2(1), 58–64.

Huang, Y. and Powell, J. W. 2012. Detecting regions of disequilibrium in taxi services under uncertainty. *Proceedings of the ACM International Conference on Advances in Geographic Information Systems, ACM-GIS*, Redondo Beach, CA, pp. 139–148.

Jiang, S., Ferreira, Jr. J. et al. 2012. Discovering urban spatial-temporal structure from human activity patterns. *Proceedings of the ACM SIGKDD International Workshop on Urban Computing, UrbComp*, Beijing, China, pp. 95–102.

Kazerani, A. and Winter, S. 2009. Can betweenness centrality explain traffic flow. *Proceedings of the 12th AGILE International Conference on GIS*, Hannover, Germany.

Leung, I. X. Y., Chan, S.-Y. et al. 2011. Intra-city urban network and traffic flow analysis from GPS mobility trace. http://arxiv.org/abs/1105.5839. Last accessed May 25, 2013.

Li, B., Zhan, D. et al. 2011. Hunting or waiting? Discovering passenger-finding strategies from a large-scale real-world taxi dataset. *IEEE International Conference on Pervasive Computing and Communications Workshops, PerComW*, Seattle, WA, pp. 63–68.

Metropolitan Transportation Authority (MTA). 2013. Subway and bus ridership. Online at http://www.mta.info/nyct/facts/ridership/. Last accessed October 29, 2013.

Port Authority of New York & New Jersey. 2013. LaGuardia airport. Online at http://www.panynj.gov/airports/laguardia.html. Last accessed October 29, 2012.

Powell, J. W., Huang, Y. et al. 2011. Towards reducing taxicab cruising time using spatio-temporal profitability maps. *Proceedings of International Symposium on Advances in Spatial and Temporal Databases, SSTD*, Minneapolis, MN, pp. 242–260.

Qi, G., Li, X. et al. 2011. Measuring social functions of city regions from large-scale taxi behaviors. *IEEE International Conference on Pervasive Computing and Communications Workshops, PerComW*, Seattle, WA, pp. 384–388.

Richter, K.-F., Schmid, F., and Laube P. 2012. Semantic trajectory compression: Representing urban movement in a nutshell. *Journal of Spatial Information Science* 4(1), 3–30.

Uddin, M. R., Ravishankar, C., and Tsotras, V. J. 2011. Finding regions of interest from trajectory data. *Proceedings of IEEE International Conference on Mobile Data Management, MDM*, Luleå, Sweden, pp. 39–48.

Wei, L.-Y., Zheng, Y., and Peng, W.-C. 2012. Constructing popular routes from uncertain trajectories. *Proceedings of the ACM International Conference on Knowledge Discovery and Data Mining, KDD*, Beijing, China, pp. 195–203.

Yan, Z., Chakraborty, D. et al. 2011. SeMiTri: A framework for semantic annotation of heterogeneous trajectories. *Proceedings the International Conference on Extending Database Technology, EDBT*, Uppsala, Sweden, pp. 259–270.

Yuan, J., Zheng, Y. et al. 2010. T-drive: Driving directions based on taxi trajectories. *Proceedings of the ACM International Conference on Advances in Geographic Information Systems, ACM-GIS*, San Jose, CA, pp. 99–108.

Yuan, J., Zheng, Y. et al. 2011a. Driving with knowledge from the physical world. *Proceedings of the ACM International Conference on Knowledge Discovery and Data Mining, KDD*, San Diego, CA, pp. 316–324.

Yuan, J., Zheng, Y. et al. 2011b. Where to find my next passenger. *Proceedings of the 13th International Conference on Ubiquitous Computing, UbiComp*, Beijing, China, pp. 109–118.

Yuan, J., Zheng, Y., and Xie, X. 2012. Discovering regions of different functions in a city using human mobility and POIs. *Proceedings of ACM International Conference on Knowledge Discovery and Data Mining, KDD,* Beijing, China, pp. 186–194.

Zhang, D., Li, N. et al. 2011. iBAT: Detecting anomalous taxi trajectories from GPS traces. *Proceedings of the 13th International Conference on Ubiquitous Computing, UbiComp,* Beijing, China, pp. 99–108.

Zhang, J. 2012. Smarter outlier detection and deeper understanding of large-scale taxi trip records: A case study of NYC. *Proceedings of the First International Workshop on Urban Computing, UrbComp,* Beijing, China, pp. 157–162.

Zhang, J., Camille, K. et al. 2012. U²SOD-DB: A database system to manage large-scale ubiquitous urban sensing origin-destination data. *Proceedings of the First International Workshop on Urban Computing, UrbComp,* Beijing, China, pp. 163–171.

Zheng, Y., Liu, Y. et al. 2011. Urban computing with taxicabs. *Proceedings of the 13th International Conference on Ubiquitous Computing, UbiComp,* pp. 89–98.

Zheng, Y. and Zhou, X. 2011. *Computing with Spatial Trajectories.* Springer, New York.

Ziebart, B. D. et al. 2008. Navigate like a cabbie: Probabilistic reasoning from observed context-aware behavior. *Proceedings of the 10th International Conference on Ubiquitous Computing, UbiComp,* Seoul, South Korea, pp. 322–331.

11 Geoinformatics and Social Media
New Big Data Challenge

Arie Croitoru, Andrew Crooks,
Jacek Radzikowski, Anthony Stefanidis,
Ranga R. Vatsavai, and Nicole Wayant

CONTENTS

11.1 INTRODUCTION: SOCIAL MEDIA AND AMBIENT GEOGRAPHIC INFORMATION

Fostered by Web 2.0, ubiquitous computing, and corresponding technological advancements, social media have become massively popular during the last decade. The term social media refers to a wide spectrum of digital interaction and information exchange platforms. Broadly, this includes blogs and microblogs (e.g., Blogger, WordPress, Twitter, Tumblr, and Weibo), social networking services (e.g., Facebook, Google+, and LinkedIn), and multimedia content sharing services (e.g., Flickr and YouTube). Regardless of the particularities of each one, these social media services share the common goal of enabling the general public to contribute, disseminate, and exchange information (Kaplan and Haenlein, 2010). And this is exactly what the general public does, making social media content a sizeable and rapidly increasing chunk of the digital universe. Facebook announced in 2012 that its system deals with petabyte scale data (InfoQ, 2012) as it processes 2.5 billion content elements and over 500 TB of data daily (TechCrunch, 2012).

This social media content is often geo-tagged, either in the form of precise coordinates of the location from where these feeds were contributed or as toponyms of these locations. Based on studies by our group using the GeoSocial

Gauge system that we developed to harvest and analyze social media content (Croitoru et al., 2012), we have observed that on average, the percentage of precisely geolocated (at the level of exact coordinates) tweets ranges typically between 0.5% and 3%. Depending on the area of study and underlying conditions, this rate may occasionally go higher. For example, a dataset collected from Japan following the Fukushima disaster reflected a data corpus where 16% of the tweets were precisely geolocated (Stefanidis et al., 2013b). This spike is attributed to the fact that the dataset from Japan reflected a technologically advanced community that was on the move (following the tsunami and subsequent nuclear accident), in which case users were tweeting using primarily their cell phones. Both of these situations, namely, the proliferation of technology in a society and an increased use of mobile (and other location aware) devices to post tweets, are conditions that tend to produce higher rates of geolocated content. Inaddition to precisely geolocated tweets, we have observed that approximately 40%–70% of tweets come with a descriptive toponym related to the location of the user. Regarding imagery and video contributed as part of social media, a recent study has indicated that approximately 4.5% of Flickr and 3% of YouTube content is geolocated (Friedland and Sommer, 2010).

The geographic content of social media feeds represents a new type of geographic information. It does not fall under the established geospatial community definitions of crowdsourcing (Fritz et al., 2009) or volunteered geographic information (VGI) (Goodchild, 2007) as it is not the product of a process through which citizens explicitly and purposefully contribute geographic information to update or expand geographic databases. Instead, the type of geographic information that can be harvested from social media feeds can be referred to as *ambient geographic information* (AGI) (Stefanidis et al., 2013b); it is embedded in the content of these feeds, often across the content of numerous entries rather than within a single one, and has to be somehow extracted. Nevertheless, it is of great importance as it communicates instantaneously information about emerging issues. At the same time, it provides an unparalleled view of the complex social networking and cultural dynamics within a society and captures the temporal evolution of the human landscape.

Accordingly, social media feeds are becoming increasingly *geosocial* in the sense that they often have a substantial geographic content. At the same time, we can observe the underlying social structure of the user community by studying the interactions among users. For example, we can identify the trail of a tweet as it is retweeted within the user community, or we can construct a social network describing the *follow* connections among numerous users. This allows us for the first time to explore the physical presence of people together with their online activities, enabling us to link the cyber and physical spaces on a massive scale. This information contributes additional content to social media (i.e., space) and provides additional context to analyze these data (i.e., topics and sentiment accord). For example, we can identify geographical hot spots of cyber communities that participate in a specific discussion and their interactions with other communities. Accordingly, geosocial analysis is inherently complex, as it comprises the study of content, connections, and locations and their variations over time. As such, it represents an emerging alternate form of geographic information, which, through its volume and richness, opens new avenues and research challenges for the understanding of dynamic events and situations.

The objective of this chapter is to present some particular challenges associated with big geosocial data, in order to provide an understanding of the corresponding analysis and processing needs. Accordingly, this chapter is organized as follows. In Section 11.2, we discuss some particular characteristics of social media as they compare to traditional big spatial data. In Section 11.3, we focus on the complexity aspect of social media content, using representative examples. In Section 11.4, we address the integration of diverse content to enable the integrative geosocial analysis of multiple social media feeds. Finally, in Section 11.5, we conclude with a view of the outlook.

11.2 CHARACTERISTICS OF BIG GEOSOCIAL DATA

A recent definition of big data (TechAmerica, 2012) is moving beyond sheer data *volume* to identify it through two additional properties, namely, *velocity* and *variety* as shown Figure 11.1. In this context, velocity refers not only to the rate at which the data is produced but also to the currency of its content and the corresponding need for timely analysis. The need to process data and extract information from them at streaming rates is imposing substantially higher computational demands than the periodic (e.g., daily or weekly) processing of comparable information (e.g., as may be the case when addressing remote sensing data). Variety on the other hand refers to the diversity of the data sources and types that are processed and on the degree to which the information to be extracted is distributed among such diverse sources. It is not uncommon for an event to be communicated by the general public in fragments, as individuals may only have a partial view of the event they are reporting. Accordingly, the event may be communicated across numerous social

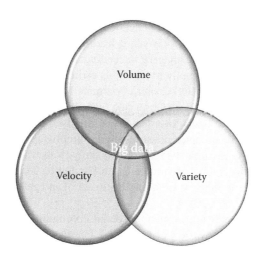

FIGURE 11.1   The three dimensions of big data: volume, velocity, and variety.

media channels by multiple users using various modalities (e.g., text in Twitter, images in Flickr and Instagram, and videos in YouTube). One of the earliest manifestations of this was during the Mumbai terrorist attacks in 2008, where Flickr imagery, Twitter streams, Google maps mashups, and Wikipedia articles were set up immediately, to provide real-time coverage of the unfolding events (Arthur, 2008). While each piece of information separately adds to our understanding of the event, the aggregate view of all these pieces offers a far better understanding of the full complexity of the actual event. This introduces the need to develop a better capacity to process *geosocial multimedia* in order to extract knowledge from diverse social media feeds. This is analogous to information aggregation in a geosensor network (Stefanidis and Nittel, 2004), where each sensor contributes a piece of information, but it is through aggregation across that the observed event is revealed in all its complexity. When it comes to social media, people act as sensors too, reporting their observations in the form of multimedia feeds, and the challenge is to compose these fragmented contributions into a bigger picture, overcoming the limitations of individual perception. Handling this variety imposes constraints on both IT architecture and algorithmic needs and will be discussed in more detail in Section 11.4. At the same time, it is important to note that the ability to monitor human observations at a massive scale and to cross-reference such data across a variety of sources and modalities (e.g., text, imagery, video, and audio) presents a unique opportunity to validate information regarding events as they unfold in space and time. We can therefore postulate that Linus's law (Raymond, 1999), as has been discussed in the context of VGI (Haklay et al., 2010; Sui and Goodchild, 2011) also has a central role in AGI and geosocial analysis.

Since their inception, geospatial datasets have always been large volume datasets, at the edge of the computational capabilities of each era. This was true at the time of early seminal computer mapping software environments in the late 1960s and early 1970s, such as SYMAP and its use in Waldo Tobler's movie of the urban expansion of Detroit (Tobler, 1970), and it is certainly true today. As a matter of fact, one could argue that geospatial data collection advances have outpaced computational improvements over the past 40+ years, leading, for example, to Digital Globe currently having an archive of imagery exceeding 4 billion km^2 in coverage and generating 1–2 PB of data annually to add to its estimated up to 30 PB of archived data. At the same time, the EROS Center is estimated to have over 4 PB of satellite imagery available, the NASA Earth Observing System Data and Information System (EOSDIS) is estimated to house over 7.5 PB of archived imagery (Maiden 2012), and NOAA's climatic data center houses a digital archive of over 6 PB. NASA is currently generating approximately 5 TB of data daily (Vatsavai and Bhaduri, 2013). Furthermore, the proliferation of Google Earth has led to Google maintaining an archive of over 20 PB of imagery, from satellite to aerial and ground-level street view imagery (*ComputerWeekly*, 2013).

Technological advances moved the geospatial community further into big data territory, by broadening the variety of geospatial datasets that are collected and analyzed and by improving the rates at which such data are generated. A Google Street View car driving down a street is equipped with multiple cameras and LIDAR sensors, capturing millions of image pixels per second, while it also simultaneously scans thousands of point locations to generate thick point clouds (Golovinskiy et al., 2009). At the same time, video surveillance is generating massive datasets using ground- or airborne-based platforms. DARPA recently unveiled ARGUS-IS (Engadget, 2013) a 1.8-gigapixel video surveillance platform that can monitor a 25 km^2 wide area at a pixel resolution of 15 cm and temporal resolution of 12 fps from an altitude of 6 km. At the same time, ground-based video surveillance has been growing extremely rapidly. As a telling reference, in the United Kingdom alone, it is estimated that between 2 and 4 million CCTVs are deployed, with over 500,000 of them operating in London (Norris et al., 2004; Gerrard and Thompson, 2011). By adding velocity and variety to big volumes, these advances have further solidified the big data nature of geospatial datasets. While these challenges are indeed substantial, they reflected an evolution rather than a breakpoint for the geoinformatics community. The objective was only partially altered: applying established analysis techniques onto larger volumes of data.

The emergence of social media however is pausing a different type of big data challenge to the geoinformatics community, due to the particularities of the analysis that these data support, a hybrid mix of spatial and social analysis, as we will see in Sections 11.3 and 11.4. At the same time, these social media datasets have some particular characteristics that differentiate them from traditional geospatial datasets. We can identify three such particular characteristics as follows:

1. *Social media datasets are streaming big data that are best-suited for real-time analysis.* As public participation in social media is increasing very rapidly, the information published through such sites is clearly exceeding big data levels. For example, in 2012, Twitter users were posting nearly 400 million tweets daily, or over 275k tweets per minute (Forbes, 2012), doubling the corresponding rates of 2011 (Twitter, 2011). At the same time, 100 million active users are uploading daily an estimated 40 millions of images in Instagram (Instagram, 2013). Furthermore, every minute, Flickr users upload in excess of 3000 images (Sapiro, 2011), and YouTube users upload approximately 72 h of video (YouTube, 2013). Accordingly, one can argue that, compared to traditional geospatial big data, social media impact more the velocity component of the diagram in Figure 11.1. Furthermore, because of their nature, social media are primarily suited to communicate information about rapidly evolving situations, ranging from civil unrest in the streets of Cairo during the Arab Spring events (Christensen, 2011) or New York during Occupy Wall Street (Wayant et al., 2012) to reporting natural disasters like a wildfire (De Longueville et al., 2009) or an earthquake (Zook et al., 2010; Crooks et al., 2013). Accordingly, not only are social media data available at streaming rates, but also their analysis must often be done at comparable rates, in order to best take advantage of the unique opportunities that they introduce.

2. *Social media data are non-curated and therefore their reliability varies substantially.* Traditional geospatial data collection campaigns are based on

meeting strict accuracy standards. This tends to produce datasets that are authoritative and fully reliable. In contrast, social media content is uncontrolled and as such displays substantial variations in accuracy. As is the case with other online media outlets, social media is not immune to the dissemination of misinformation. In Figure 11.2, we show, for example, how after the Boston Marathon bomb attacks the wrong names of suspects were circulated in Twitter traffic for few hours before the real suspects were identified. Nevertheless, studies indicate that information dissemination patterns can be analyzed to assess the likelihood that a trending story may be deemed likely to be true or not (Castillo et al., 2011). Furthermore, the massive membership of social media is offering another mechanism to verify breaking stories. As we see in Figure 11.3, real events reported from the ground during a demonstration tend to be clustered in space and time, making it easier to verify content reliability (Croitoru et al., 2011). These two examples demonstrate quite vividly both the unique suitability of social media to report breaking stories, as well as the variability in accuracy of their content, which are both distinguishing them from traditional data sources.

3. *The spatial distribution of social media contributions is nonuniform and heavily skewed.* Traditional geospatial data offer broad and consistent coverage: satellites provide global remote sensing imagery coverage with high revisit frequency (from hours to few days), and numerous local campaigns complement that foundation by providing additional data. Social media contributions do not follow this pattern though. Their content reflects the participation and contributions of a particular demographic. Internet users under the age of 50 are mostly involved with them (with those in the 18–29 age group slightly leading the 30–49 group in terms of participation), college graduates participating more than nongraduates and women also slightly leading men (Duggan and Brenner, 2012). Each particular service has its own profile, but they all tend to fall under these general participation patterns. The users are also living primarily in urban settings, and this is affecting the spatial coverage of these feeds.

Social media contributions display a spatial distribution that reflects the aggregation of the previous trends. They tend to initiate more from urban, technologically advanced areas, rather than rural remote locations. Figure 11.4 is quite representative of this pattern of predominantly urban contributions, with tweets in the broader New York area reporting the Virginia earthquake in the summer of 2011. We can see that the reports are most heavily contributed from the Manhattan region, and their density is dropping substantially as we move away toward suburbs and the less urban regions in New Jersey. Nevertheless, these contributions also often originate from locations of interest to the general public. Figure 11.5 shows a sample of the variety of imagery that has been contributed to Flickr from the Chernobyl area (site of the 1986 nuclear disaster).

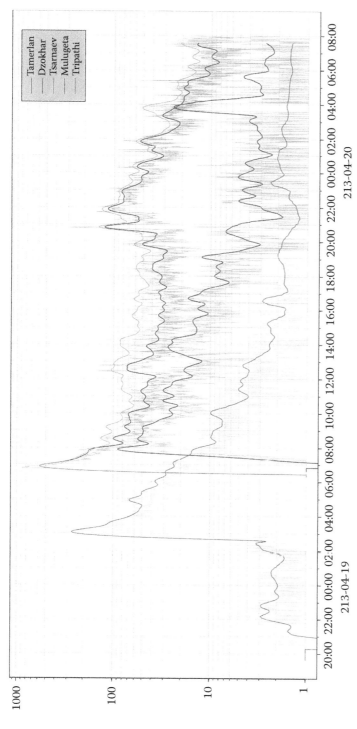

FIGURE 11.2 A graph showing how Twitter users reported the names of the wrong suspects following the Boston Marathon bombing of April 15, 2013. The horizontal axis is time of the day, extending from the evening of the 18th (leftmost point) to the morning of the 20th (rightmost point). The vertical axis represents the log of tweet mentions for each suspect name. The long-dashed and dash-dot lines correspond to the names of two wrongly accused individuals (Tripathi and Mulugeta respectively), while the dotted, dashed, and solid lines are the last (Tsarnaev) and first (Dzokhar, dashed; Tamerlan, solid) names of the actual suspects.

FIGURE 11.3 Clustered geolocated tweets (black dots) and *Flickr* imagery (camera icons) reporting events as part of Occupy Wall Street's Day of Action on November 17, 2011. (From Croitoru A. et al., Towards a collaborative GeoSocial analysis workbench, *COM-Geo*, Washington, DC, 2012.)

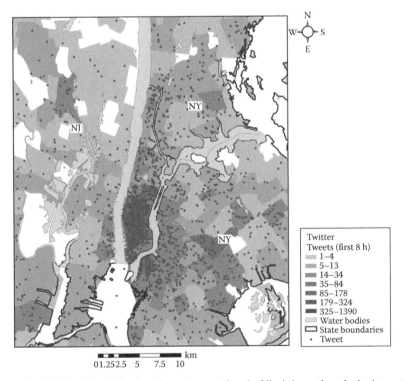

FIGURE 11.4 Spatial distribution of tweets reporting the Virginia earthquake in August 23, 2011. (From Crooks A.T. et al., *Trans. GIS*, 17(1), 124, 2013.)

FIGURE 11.5   Spatial distribution of Flickr imagery from the Chernobyl area.

11.3 GEOSOCIAL COMPLEXITY

In addition to the particular characteristics that we identified previously in Section 11.2, geosocial data are also differentiated from traditional geospatial datasets due to their complexity. More specifically, they are predominantly linked information; links are established among users to establish a social network and among words to define a storyline that is communicated through pieces of information. Regarding user connections, they can be established through specific actions. For example, *user A* following, replying, referring to, or retweeting *user B* can establish a connection among them. Aggregating all these connections provides us with a view of the users as a structured, networked community that can be represented as a graph.

In Figure 11.6, we show a sample network, generated by considering retweet activities. The network was constructed using Twitter data discussing the 2012 US presidential elections. Nodes (dots) correspond to users, and lines connecting them indicate instances where a user retweeted another. This graph representation offers detailed insights into the structure of a social network. We can identify, for example, a bigger cluster of nodes on the left, formed around nodes 687, 685, 667, and 642, and their connected groups. These groups are interconnected among themselves to form a larger cluster, in which these four nodes play a leading role (as they lead the corresponding smaller groups). We have used different colors for each group to better communicate individual group structure and the points where two groups connect to form a larger community. On the right side of the same figure, we identify a smaller cluster, formed around nodes 1046, 1044, and 805 (and their corresponding groups), that remains disjoint from the left cluster. We also observe smaller disconnected

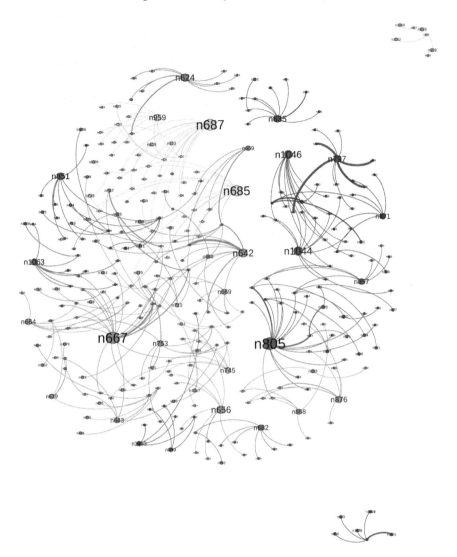

FIGURE 11.6 A sample social network structure.

group formed around node 635 (blue cluster at the upper middle) and some other further disjoint small clusters at the top and bottom of that figure.

Similarly to users, words are also linked, as they appear in the same messages, to form word networks that define the discussion themes. In Figure 11.7, we show a snapshot of such a word network. The size of the node corresponds to the frequency of this word, and we can see how *obama* is a keyword that is emerging from the discussion, followed by *romnei, bain, barack, patriot,* etc. Connections among words are established every time two words appear together in a single record (e.g., a tweet or the caption of an image), and the thickness of the edge connecting two words in this network is proportional to the frequency of the two words appeared together in the same tweet. It should be noted that in this word network, we use the stemmed

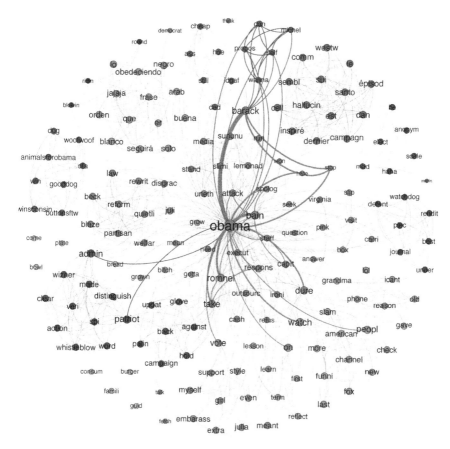

FIGURE 11.7   A word network structure corresponding to the data corpus of Figure 11.6.

form of words, a process that reduces (normalizes) words to a common root form to reduce the effect of form variations on the word frequency count. For example, the word *Romney* would be stemmed to the form *romnei*, and the word *people* would be stemmed to the form *peopl*. A detailed review of stemming algorithms can be found in (Hall, 1996).

While we tried to use a small data corpus to visualize these networks, it is clear to see that they can become very complicated very fast. A collection of tweets captured during Hurricane Sandy (October 2012) resulted in a data corpus of nearly 10 million tweets, including over 4.7 million retweets (47.3% of the data corpus) and over 4.1 million (41.7% of the data corpus) links to external sites (e.g., Flickr photos, news websites). We can understand that analyzing such highly complex datasets is a computational challenge, which is typically addressed through graph analysis. Enumerating interesting patterns, for example, triangles in a large network, is a computationally expensive task. This task requires a listing algorithm, which requires at least one operation per triangle, leading to a worst-case lower computational bound of $\Omega(n^3)$.

A variety of solutions exist to support feature extraction from such network graphs by constructing (or extracting) variables from them. These features (variables) typically encode the position and importance of each node with respect to the other nodes in the graph. Centrality, topological, and distance measures are fundamental metrics to support such analysis:

- *Centrality metrics*: Centrality is one of the most important structural attributes of a social network (Freeman, 1979). Often it is defined in terms of *reachability* of a given node in a network. Some of the centrality measures can also be defined in terms of *flow*. In recent years, a great many measure have been proposed, including degree, closeness, betweenness, and flow betweenness (Scott, 2012). Degree centrality refers to the network activity for a node, often measured as the number of connections a node has. The betweenness measure reflects the influence of a node in the network, often measured as the number of times a node acts as a bridge along the shortest path between two other nodes (Wasserman and Faust, 1994). The closeness measure refers to the visibility of a node. Closeness can be extracted in terms of geodesic distances or eigenvector measure. PageRank (Page et al., 1999) and the Katz centrality (Katz, 1953; Borgatti, 2005) are two other measures that are closely related to the eigenvector centrality measure. Computing these features on large graphs is superlinearly expensive, and scalability of these algorithms in terms of three graph partitioning methods and GPU implementation were studied by Sharma et al. (2011). Some of the centrality measures like betweenness (shortest path or random walk) are computationally expensive $-O(n^3)$. It is noted in Kang et al. (2011) that parallelizing (graph partitioning) these measures is not straightforward. Therefore, implementing these measures in main memory systems like uRiKA is an important research task.
- *Topological features*: Topological features are dependent on abstract structure of the graph and not on the representation. Commonly used features include degree distribution, clustering coefficient, and connectivity. The degree distribution is probability distribution of the degree (of nodes) over the entire network. It is an important measure, as random graphs often have binomial or Poison distributions, whereas real-world networks are highly skewed (Bollobas et al., 2001), that is, the majority of their nodes have low degree but a small number of them (also known as hubs) have high degree. In scale-free networks, the degree distribution follows a power law. The clustering coefficient determines the degree to which the nodes tend to cluster together (Huang, 2006). Another reliable measure is the effective diameter, which is measured as the minimum number of hops in which some fraction of all connected pairs of nodes can reach each other.
- *Similarity and distance measures*: Searching for similar patterns in heterogeneous information networks is a challenging task. Heterogeneous networks are directed graphs, which contain multiple types of objects or links. There exist several similarity measures for networks with same type of nodes or links. However, the same is not true for heterogeneous networks.

Recently, several new similarity measures have been defined for heterogeneous networks (Sun and Han, 2012). A particular similarity measure, called PathSim (Sun et al., 2011b), was shown to find many meaningful objects as compared to the random-walk-based similarity measures. Authors have also provided prediction and clustering algorithms using this similarity measure.

11.4 MODELING AND ANALYZING GEOSOCIAL MULTIMEDIA: HETEROGENEITY AND INTEGRATION

Currently, social media services offer a wide range of platforms using various technologies and platforms. As a result, their content tends to be very diverse both in content—ranging from text to photos and videos—and in form, ranging from structured content to semi- or nonstructured content. In addition, the form of raw social media data tends to be unstructured or ill-defined, making valuable knowledge hidden and limiting the capability to process it through automation (Sahito et al., 2011). For example, both Twitter and Flickr provide application programming interfaces (APIs) to query their content, but their responses are often structurally incompatible, not only between services but also within a single service. For example, the Twitter API can return the same content in different formats depending on the particular API calls made. Managing and integrating such diverse social media data requires the development of a unified conceptual data model that will support the various data structures under a single scheme. Generally, this task can be viewed as a data-cleaning problem, that is, the removal of errors and inconsistencies in databases (Rahm and Do, 2000), from either a single source or multiple sources of data. For multiple data sources, data-cleaning problems can arise at the schema level (e.g., structural inconsistencies) or at the instance level (e.g., uniqueness violations).

A step toward a more general solution for integrating social data was recently presented by Lyons and Lessard (2012), who introduced a social feature integration technique for existing information systems that are not socially oriented. However, there seem to be no widely established universal conceptual models that could be directly applied to multiple social media sources for geographical analysis. For example, Sahito et al. (2011) presented a framework for enriching and deriving linkages in Twitter data by using semantic web resources, such as DBpedia* and GeoNames.† However, this work considers only a single social media source, that is, Twitter. A more generalized conceptual model has been recently introduced by Reinhardt et al. (2010), in which two networks are considered: a set of artifact networks that describes the relationships between data elements (e.g., chats, blogs, or wiki articles) and a social network. This model, Artefact-Actor-Network (AAN), is created by linking the two networks through semantic relationships. Another closely related work is the data model presented by Shimojo et al. (2010), which focuses on lifelogs: digital records of the experiences and events a person encounters during a period of time, which are generated by individuals

---

* http://dbpedia.org/
† http://www.geonames.org/

(Kalnikaitė et al., 2010). Their work presents a model that is geared toward the integration of multiple heterogeneous social media sources through the introduction of a common model for lifelog data.

Building on this prior work, we present here a data model for converting social media data into structured geosocial information, from which knowledge can be extracted through further analysis. Focusing on Twitter (text) and Flickr (imagery), Figure 11.8 depicts an entity–relation (ER) diagram of our data model. It is comprised

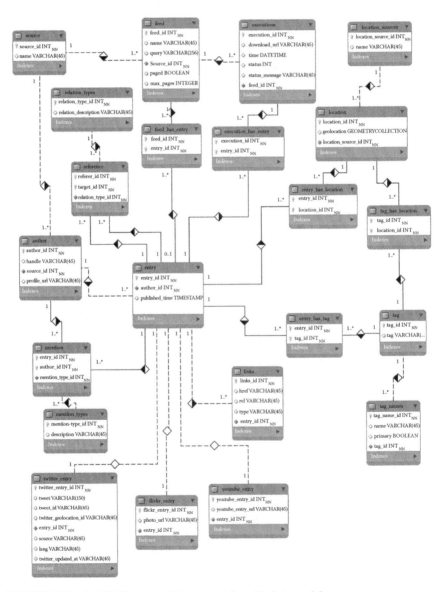

FIGURE 11.8   An ER diagram of the geosocial media data model.

of several primary building blocks, namely, entry, geolocation, time, keywords, and authors. As some of these components may have many-to-many relationships between them, additional tables were added to represents such relationships. In the following discussion, we describe these components and the relations between them:

- *Entry*: An information entry serves as an abstract record of social media data, a high-level entity that binds the various components that constitute a data entry. An entry instance is uniquely identified and is linked to one author instance of the author component. In addition, each entry is associated with a time stamp indicating when was the entry instance was published.
- *Source API*: The source API data are comprised of the data elements that are unique to each social media source. For example, the Twitter API returns a tweet attribute that contains the content of a tweet, while the Flickr API returns a photo URL and has no equivalent to the tweet attribute. Such source-specific attributes, which are driven by the characteristics of each social media source, are regarded as source dependent and are therefore stored separately in dedicated tables.
- *Author*: Author instances represent social media service users that contributed content. As user identification across sources is rather limited, each social media service creates its own author namespace, that is, a unique set of user names. As an author is identified by a tuple of a user name and a social media service identifier, different users can have the same identifier value in different services. It should be noted that authors can be referenced to in the content of social media feeds, through which the underlying social network can be recovered. In our model, this is accomplished by linking entries to users through the *mentions* table.
- *Geolocation*: As discussed earlier, geolocation information for social media feeds can be inferred *indirectly* from content analysis, or it can be extracted *directly* from the data itself. The contributors themselves may directly provide geolocation information, either in the form of exact coordinates or as a toponym (e.g., listing a city name) that in turn can be geolocated using any gazetteer service (e.g., Lieberman et al., 2010). A more detailed discussion on the various forms of this geolocation content can be found in Marcus et al. (2011), Croitoru et al. (2012), and Crooks et al. (2013). In our data model, geolocation information is stored in the form of coordinates, alongside information about the source through which they were acquired.
- *Time*: Temporal information can typically be found in all social media platforms. In the case of Twitter, temporal information is provided in the form of a time stamp of submission time (instance when it was submitted to Twitter by the user). In Flickr, the time tag can represent three different types of time attributes: posted time (the actual upload time), taken time (the time stamp embedded in the image metadata), and last update time (the last moment that the description of the picture in *Flickr* has been updated). In the data model, time information is embedded with each entry instance along with a time stamp-type identifier.

- *Keywords*: As part of a social media entry, users contribute keywords or tags, like hashtags (#) in Twitter (Huang et al., 2010) or user tags in Flickr (Ames and Naaman, 2007) to emphasize views and ideas and engage other users (Romero et al., 2011). Hashtag usage, for example, has been shown to accelerate data retrieval from Twitter (Zhao et al., 2011) and Flickr (Sun et al., 2011a). Hashtags also support the building of semantic networks by allowing individual tweets to be linked thematically based on their content. Unlike explicit tagging, implicit keyword may emerge from user conversations, when certain words become widely adopted as a result of a noteworthy event. In the data model, unique keyword instances are stored in a separate tags table, which is linked to the *entries* table.
- *Source*: The social media source component (and its corresponding *sources* table) serves as a directory of the different source types that can provide data to the analysis framework. Each source can be associated with one or more feeds, which are instances of a given source type. As there may be situations where a single entry is present in multiple feeds (e.g., an identical tweet that is retrieved from two different feeds using different queries), only a single copy of the tweet is stored and a reference to the feed is made to avoid data redundancy along with a unique identifier.

Based on this integrated geosocial multimedia data model, an analysis framework can be designed. The analysis of geosocial multimedia data requires a novel analysis framework that expands the capabilities of traditional spatial analysis to account for the unique components of social data, for example, social links and non-geographic content (Wayant et al., 2012). Toward this goal, we present a framework for collaborative geosocial analysis, which is designed around data harvesting from social media feeds and a collaborative analysis workbench geared toward producing actionable knowledge. This framework is designed to enable and support distributed synthesis, knowledge production and information dissemination by communities of subject matter experts (SMEs), and other end users. To support such activities, the framework is designed around four primary themes: data harvesting and gathering; analysis tools including spatial, temporal, and social analyses; support for SME communities interaction; and support of end-user communities that process data and interact and provide feedback. Each of these components should be modular and extensible, to allow the framework to adapt to a wide range of problem domains, for example, a new social media feed can be easily added without a need to redesign existing data handling and functionality capabilities.

An overview of this framework is shown in Figure 11.9. The process starts by gathering data from social multimedia, using, for example, specific regions of interest (ROIs), time intervals, and user-provided keywords. The harvested social media data are then analyzed and disseminated to the user community. Analysis results can then be used to refine the data gathering process by refining, for example, the ROIs, time intervals, or the keywords used for the analysis. Social multimedia is therefore not regarded simply as an additional data source in this process, but rather as a primary driver for the entire analysis and knowledge generation workflow.

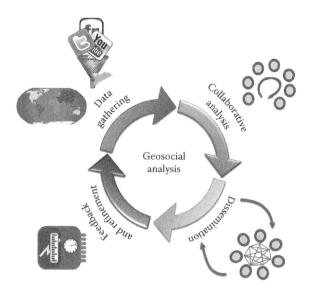

FIGURE 11.9   The geosocial analysis framework.

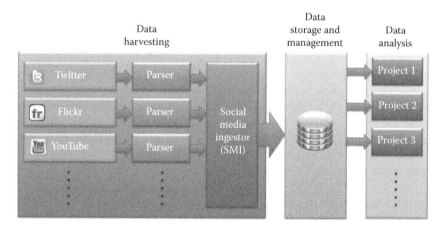

FIGURE 11.10   A high-level architecture of a geosocial analysis system.

The continuous refinement process is dependent upon the ability to effectively collect and store social multimedia (as was outlined in the geosocial multimedia model). A general architecture of a system to collect data and extract geosocial information from multiple social media feeds comprises three components (Figure 11.10): data collection from the corresponding social media data providers via APIs; data parsing, integrating, and storing in a resident database; and data analysis to extract information of interest from them.

Once data are collected and processed through the social media ingestor, analysis of the data can take place through a collaborative workbench (Figure 11.11). An underlying design goal of this workbench framework is to foster knowledge discovery and support the collaboration of SME communities through a collaborative web-based

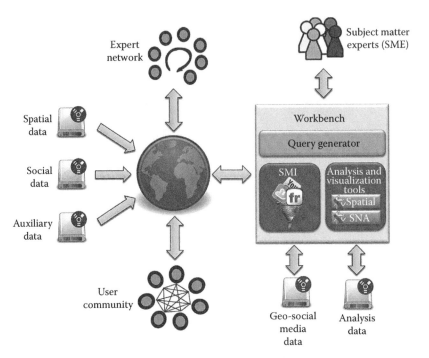

FIGURE 11.11 A schematic view of the geosocial analysis workbench.

platform. Using such platform, the collection, aggregation, analysis, and synthesis of geosocial data can be carried out and disseminated to geographically distributed user communities. At the same time, end-user communities are also empowered to interact with SME communities, to respond and contribute insights and knowledge by providing feedback and inputs to the analysis process. To accomplish this design goal, the workbench is designed around several interrelated enablement principles (Croitoru and Arazy, 2007):

- *Spatial enablement*—the capacity to represent and reference events in a real-world spatial setting. Representation includes the ability to model data for different spatial entities, for example, points, lines, and polygons, as well as other high-level representations of events, especially spatiotemporal (see, e.g., Laube et al., 2007). Such enablement also includes the ability to store, query, and retrieve spatial data and support auxiliary thematic layers (e.g., census data).
- *Temporal enablement*—the capacity to associate data entities with a time stamp. Time is essential not only for understanding when things happen but also for evaluating data relevancy. In addition, temporal enablement allows users to make predictions through modeling, thus enabling users to combine historical observations and future predictions in their analysis.
- *Social media enablement*—the capacity to harness social media feeds for geosocial analysis (e.g., Twitter *or* Flickr) and issue data collection requests. The collection mechanism provides the capability to harvest social media

based on location and time, as well as based on thematic keywords (or tags), therefore supporting a theme-based query mechanism for social media.

- *Collaboration enablement*—the capacity to empower groups of SMEs and user communities to collaboratively forge a data corpus for addressing a common goal and then analyze it toward the production of actionable knowledge. Such enablement includes the ability to collaboratively isolate and gather relevant data and process and analyze it toward the production of knowledge.
- *Geosocial analytics enablement*—the capacity to embed social media in a spatiotemporal domain for aggregation, analysis, and derivation of analysis products, both qualitative and quantitative. Such tools are provided to users in the form of thematic toolboxes, for example, a social network analysis (SNA) toolbox, a spatial analysis toolbox, and simulation and a prediction analysis toolbox. Based on these toolboxes, users can create and share subject matter workflows for processing social media data, thus allowing experts to collaboratively perform analysis. It is worth noting that the workbench also enables the creation of subject matter toolboxes in order to encapsulate domain-specific expert knowledge into analysis tools.

11.5 OUTLOOK: GRAND CHALLENGES AND OPPORTUNITIES FOR BIG GEOSOCIAL DATA

The motivation for this chapter comes from the unprecedented developments in social media and the resulting effects on actively and passively contributed geographic information as we discussed in Section 11.1. Big data, especially big geosocial data, are only expected to become more *bigger* as time progresses. We have already seen this occur with the proliferation of smartphones and other location-aware devices, and we do not expect it to stop in the near future. We would argue that big geosocial data provide us with unique opportunities to collect real-time data in epic scale and geolocate this information for analysis. Through geosocial analysis, we can gain a greater understanding of various parameters that shape the human landscape. For example, people's tweets can act as sensor with respect to earthquakes (e.g., Crooks et al., 2013) or highlight potential hot spot emergence of political events (e.g., Stefanidis et al., 2013a). Notwithstanding its application to various other avenues of human activity including languages (e.g., Mocanu et al., 2013), elections (e.g., Tumasjan et al., 2011), riots (e.g., Tonkin et al., 2012), emotions (e.g., Quercia et al., 2012), happiness (e.g., Bliss et al., 2012), college aspirations (e.g., Wohn et al., 2013), disease spreads (e.g., Achrekar et al., 2011), and flooding (e.g., Graham et al., 2012), one could consider these as altering the notions of how we explore geographical and social systems. For example, we can observe the collapse of a physical infrastructure as it is affecting its people (e.g., Sakaki et al., 2010) or the collapse of a social system while leaving the physical infrastructure intact such as in some cases of the Arab Spring (e.g., Howard et al., 2011). In a sense, such data streams harvested from humans acting as sensors have similarities to how one uses rain and stream gauges to monitor flooding in early warning systems.

Big geosocial data therefore offer us a new lens to study human systems. Traditional geographical analysis of human systems often focused on large-scale studies of population aggregates, remotely sensed imagery, or on small-scale studies of small groups of individuals. This was dictated by data availability limitations, as census, satellites, and individual interviews were the only practical means available to collect information on societal aspects and their distribution in space. These limitations kept us from getting a fine resolution and extensive coverage view of societal issues, problems, trends, etc. Similar to the progress that the invention of the microscope brought to biology or the telescope to astronomy, we are now witnessing a comparable paradigm shift in our ability to observe and analyze sociocultural expressions in space. Our ability to harvest geolocated social media feeds and crowdsourced data is offering unprecedented opportunities to monitor people's actions, reactions, and interactions at a unique combination of fine spatiotemporal resolution and massive coverage.

But just like the invention of the microscope that introduced new types and volumes of data in biology and fostered the emergence of novel analysis approaches and microbiology, so too the emergence of social media and crowdsourced data presents the need for a new geosocial analysis paradigm as was discussed in Sections 11.2 through 11.4. For the first time, we can bridge the gap between the aforementioned large-scale and small-scale analyses, and we can begin to understand intrinsic connections between individuals, groups, and their environments based on sociocultural similarities. The discovery through novel social network and spatial analysis techniques of up-to-now invisible sociocultural patterns in space will advance our understanding of how our increasingly networked society is shaped, operates, and evolves in space. This is particularly applicable to urban areas, as they are home to the vast majority of the world's population; they are also home to the majority of the geosocial media crowd, and furthermore, their urban form and infrastructure define to a great extent activities and behaviors of the individuals. This urban setting will serve as the stage for our analysis.

However, there is also a growing discussion that big data need a *big theory*. For example, West (2013) argues we need a deeper understanding of complexity science, in the sense that we are looking at a system of many parts that can interact in many different ways that adapts and evolves through space and time. Big data offer us a way to observe these interactions, but we do not have a unified framework for addressing questions of complexity. To fully address this issue, we also need to link cross-disciplinary expertise that will foster the further development of transdisciplinary knowledge. For example, by linking SNA with geographical analysis, using information harvested from geosocial media feeds and crowdsourced data, we gain a new understanding of space as it is defined by and in turn defines the interaction of people within it. For the first time, we can test notions such as Tobler's (1970) first law of geography that "everything is related to everything else, but near things are more related than distant things." This has already been shown to be valid with respect to Wikipedia spatial articles (Hecht and Moxley, 2009), but is it true for other aspects of geosocial media? We can also explore on a massive scale time (time–space) geography that has seen of much research since the 1960s (Hagerstrand, 1967).

However, even with the growth of geosocial media, one has to note that there is an issue of only getting a sample of the population when collecting information from geosocial media feeds, namely, individuals who are active in this arena. Nevertheless, this sample is rapidly growing, as relevant technology adoption is becoming more ubiquitous (see Smith, 2011; Nielsen, 2012). That said, however, researchers need to consider new measures of engagement (e.g., contrasting rates with say Internet usage) along with other normalization methods (e.g., Stefanidis et al., 2013a) when looking at results from geosocial data. Coinciding with participation rates are issues relating to false information that can range from state-sponsored propaganda such as the *50-cent party* that is supported by the Chinese government and pays for blog or message-board postings that promote the Chinese Communists Party line in an attempt to promote public opinion their way (see Bremmer, 2010) to that of spam and falsification of events (e.g., Gupta et al., 2013) that are active areas of research in the data mining community.

This rise in geosocial media and the ability for analysis also raises several concerns with respect to the suitability of traditional mapping and GIS solutions to handle this type of information. We no longer map just buildings and infrastructure, but we can now map abstract concepts like the flow of information in a society and contextual information to place and link both quantitative and qualitative analyses in human geography. In a sense, one could consider AGI to be addressing the fact that the human social system is a constantly evolving complex organism where people's roles and activities are adapting to changing conditions and affect events in space and time. By moving beyond simple mashups of social media feeds to actual analysis of their content, we gain valuable insight into this complex system. What is to come next is difficult to predict. For example, consider that only 10 years ago the idea of location-based services and GPS embedded into mobile devices was still in its infancy. Advances in tools and software made geographic information gathering easier, resulting into growing trends in crowdsourcing geographical data rather than using authoritative sources (such as national mapping agencies). More recently, the popularity of geographically tagged social media is facilitating the emergence of location as a commodity that can be used in organizing content, planning activities, and delivering services. We expect this trend to increase as mobile devices become more location aware. One could relate this to the growing usage of online activities and services (such as real-time information on geosocial media sites like Foursquare, Facebook places, Google Latitude, Twitter, and Gowalla and a host of new sites and services emerging with time). But also more static sites (in the sense, one can upload when wants), such as Flickr and YouTube, provide means of viewing and in a sense forming an opinion of a place without actually visiting.

There is also the issue of privacy with respect to geosocial media. Nobody wants to live in Orwell's version of *Big Brother*; however, harvesting ambient information brings forward novel challenges to the issue of privacy, as analysis can reveal information that the contributor did not explicitly communicate (see Friedland and Sommer, 2010). But this is not a new trend; it has actually been happening for a while now (e.g., *New York Times*, 2011). Google itself is basically a marketing tool using the information it collects to improve its customer service. Similarly, Twitter makes money by licensing their tweet fire hose to search engines, while companies can pay for *promoted tweets* (see Financial Times, 2010). And this trend has already spread to locational information. For example, TomTom (2011) has been using passively sensed data for helping police with placing speed cameras. This comes alongside Apple's iPhones storing locational data while the user is unaware (*BBC* 2011). However, people are making progress in highlighting the issue of privacy relinquishing when sharing locational information. Sites and apps such as pleaserobme.com or the creepy (http://ilektrojohn.github.io/creepy/), a geolocation aggregator, have demonstrated the potential for aggregating social media to pinpoint user locations. Trying to protect people's identities in times of unrest is also a well-recognized concern, for example, the Stand by Task Force (2011) suggests ways of limiting expose and delay information for the recent unrest in North Africa. But the power of harvesting AGI stems from gaining a deeper understanding of groups rather than looking at specific individuals. As the popularity of social media is growing exponentially, we are presented with unique opportunities to identify and understand information dissemination mechanisms and patterns of activity in both the geographical and social dimensions, allowing us to optimize responses to specific events, while the identification of hot spot emergence helps us allocate resources to meet forthcoming needs.

REFERENCES

Achrekar, H., Gandhe, A., Lazarus, R., Yu, S.H., and Liu, B. (2011) Predicting flu trends using twitter data, *2011 IEEE Conference on Computer Communications Workshops (INFOCOM WKSHPS)*, Shanghai, China, pp. 702–707.

Ames, M. and Naaman, M. (2007) Why we tag: Motivations for annotation in mobile and online media, *Proceedings ACM SIGCHI Conference on Human Factors in Computing Systems*, San Jose, CA, pp. 971–980.

Arthur, C. (2008) How Twitter and Flickr recorded the Mumbai terror attacks, *The Guardian*, Available at http://bit.ly/MIMz (accessed on June 3, 2013).

BBC (2011) iPhone tracks users' movements, Available at http://bbc.in/e81tgt (accessed on June 3, 2013).

Bliss, C.A., Kloumann, I.M., Harris, K.D., Danforth, C.M., and Dodds, P.S. (2012) Twitter reciprocal reply networks exhibit assortativity with respect to happiness, *Journal of Computational Science*, 3(5): 388–397.

Bollobas, B., Riordan, O., Spencer, J., and Tusnady, G. (2001) The degree sequence of a scale-free random graph process, *Random Structures & Algorithms*, 18(3): 279–290.

Borgatti, S.P. (2005) Centrality and network flow, *Social Networks*, 27: 55–71.

Bremmer, I. (2010) Democracy in cyberspace: What information technology can and cannot do, *Foreign Affairs*, 89(6): 86–92.

Castillo, C., Mendoza, M., and Poblete, B. (2011) Information credibility on Twitter, *Proceedings of the 20th International Conference on World Wide Web*, Hyderabad, India, pp. 675–684.

Christensen, C. (2011) Twitter revolutions? Addressing social media and dissent, *The Communication Review*, 14(3): 155–157.

ComputerWeekly (2013) What does a petabyte look like? Available at http://bit.ly/Ywtnqg (accessed on October 29, 2013).

Croitoru, A. and Arazy, O. (2007) Introducing location-based mass collaboration systems, *Proceeding of the 17th Workshop on Information Technologies and Systems,* Montreal, Quebec, Canada, pp. 37–42.

Croitoru, A., Stefanidis, A., Radzikowski, J., Crooks, A.T., Stahl, J., and Wayant, N. (2012) Towards a collaborative GeoSocial analysis workbench, *COM-Geo*, Washington, DC.

Crooks, A.T., Croitoru, A., Stefanidis, A., and Radzikowski, J. (2013) #Earthquake: Twitter as a distributed sensor system, *Transactions in GIS*, 17(1): 124–147.

De Longueville, B., Smith, R.S., and Luraschi, G. (2009) OMG, From Here, I Can See The Flames!: A use case of mining location based social networks to acquire spatio-temporal data on forest fires, *Proceedings of the 2009 International Workshop on Location Based Social Networks*, Seattle, WA, pp. 73–80.

Duggan, M. and Brenner, J. (2012) *The Demographics of Social Media Users—2012*, Washington, DC: Pew Research Center, Available at http://bit.ly/XORHo0.

Engadget (2013) DARPA's 1.8-gigapixel cam touts surveillance from 20,000 feet (video), Available at http://engt.co/UyGFUF (accessed on October 29, 2013).

Financial Times (2010) Coke sees 'phenomenal' result from twitter ads, *Financial Times*, June 25, Available at http://on.ft.com/13jvSkp (accessed on June 3, 2013).

Forbes (2012) Twitter's Dick Costolo: Twitter mobile ad revenue beats desktop on some days, Available at http://onforb.es/KgTWYP (accessed on January 19, 2013).

Freeman, L.C. (1979) Centrality in social networks conceptual clarification, *Social Networks*, 1(3): 215–239.

Friedland, G. and Sommer, R. (2010) Cybercasing the joint: On the privacy implications of geotagging, *Proceedings of the Fifth USENIX Workshop on Hot Topics in Security (HotSec 10)*, Washington, DC.

Fritz, S., MacCallum, I., Schill, C., Perger, C., Grillmayer, R., Achard, F., Kraxner, F., and Obersteiner, M. (2009) Geo-wiki.org: The use of crowdsourcing to improve global land cover, *Remote Sensing*, 1(3): 345–354.

Gerrard, G. and Thompson, R. (2011) Two million cameras in the UK, *CCTV Image*, 42: 10–12.

Golovinskiy, A., Kim, V.G., and Funkhouser, T. (2009) Shape-based recognition of 3D point clouds in urban environments, *Proceedings of IEEE International Conference on Computer Vision*, Kyoto, Japan, pp. 2154–2161.

Goodchild, M.F. (2007) Citizens as sensors: The world of volunteered geography, *GeoJournal*, 69(4): 211–221.

Graham, M., Poorthuis, A., and Zook, M. (2012) Digital trails of the UK floods—How well do tweets match observations? *The Guardian*, Available at http://bit.ly/TnOvwu (accessed on May 12, 2013).

Gupta, A., Lamba, H., Kumaraguru, P., and Joshi, A. (2013) Faking sandy: Characterizing and identifying fake images on twitter during hurricane sandy, *Second International Workshop on Privacy and Security in Online Social Media*, Rio de Janeiro, Brazil.

Hagerstrand, T. (1967) *Innovation Diffusion as a Spatial Process*, Chicago, IL: The University of Chicago Press.

Haklay, M., Basiouka, S., Antoniou, V., and Ather, A. (2010) How many volunteers does it take to map an area well? The validity of Linus' law to volunteered geographic information, *The Cartographic Journal*, 47(4): 315–322.

Hall, D. (1996) Stemming algorithms—A case study for detailed evaluation, *Journal of the American Society for Information Science*, 47(1): 70–84.

Hecht, B. and Moxley, E. (2009) Terabytes of Tobler: Evaluating the first law in a massive, domain-neutral representation of world knowledge, in Hornsby, K.S., Claramunt, C., Denis, M., and Ligozat, G. (eds.), *Spatial Information Theory: Proceedings of the Ninth International Conference, COSIT 2009*, Berlin, Germany: Springer, pp. 88–105.

Howard, P.N., Duffy, A., Freelon, D., Hussain, M., Mari, W., and Mazaid, M. (2011) Opening closed regimes: What was the role of social media during the Arab spring? Working Paper 2011.1. Project on Information Technology and Political Islam, University of Washington's Department of Communication, Seattle, WA.

Huang, J., Thornton, K., and Efthimiadis, E. (2010) Conversational tagging in Twitter, *Proceedings of the 21st ACM Conference on Hypertext and Hypermedia*, Toronto, Ontario, Canada, pp. 173–178.

Huang, Z. (2006) Link prediction based on graph topology: The predictive value of the generalized clustering coefficient, *Workshop on Link Analysis: Dynamics and Static of Large Networks* (*LinkKDD2006*), Philadelphia, PA.

InfoQ (2012) Petabyte scale data at Facebook, Available at http://www.infoq.com/presentations/Data-Facebook (accessed on October 29, 2013).

Instagram (2013) Instagram stats, Available at http://instagram.com/press/ (accessed on October 29, 2013).

Kalnikaitė, V., Sellen, A., Whittaker, S., and Kirk, D. (2010) Now let me see where I was: Understanding how lifelogs mediate memory, *Proceedings of the 28th International Conference on Human Factors in Computing Systems*, Atlanta, GA, pp. 2045–2054.

Kang, U., Papadimitriou, S., Sun, J., and Tong, H. (2011) Centralities in large networks: Algorithms and observations, *SIAM International Conference on Data Mining*, Mesa, AZ, pp. 119–130.

Kaplan, A.M. and Haenlein, M. (2010) Users of the world unite! The challenges and opportunities of social media, *Business Horizons*, 53(1): 59–68.

Katz, L. (1953) A new index derived from sociometric data analysis, *Psychometrika*, 18: 39–43.

Laube, P., Dennis, T., Forer, P., and Walker, M. (2007) Movement beyond the snapshot—Dynamic analysis of geospatial lifelines, *Computers, Environment and Urban Systems*, 31(5): 481–501.

Lieberman, M.D., Samet, H., and Sankaranarayanan, J. (2010) Geotagging with local Lexicons to build indexes for textually-specified spatial data, *Proceedings of the IEEE 26th International Conference on Data Engineering*, Long Beach, CA, pp. 201–212.

Lyons, K. and Lessard, L. (2012) S-FIT: A technique for integrating social features in existing information systems, *Proceedings of the 2012 iConference*, New York, pp. 263–270.

Maiden, M. (2012) NASA Earth Science Data Systems Program, Southeastern Universities Research Association (SURA) Information Technology Committee Meeting, Washington, DC, Available at http://www.sura.org/.

Marcus, A., Bernstein, M.S., Badar, O., Karger, D., Madden, S., and Miller, R.C. (2011) Processing and visualizing the data in tweets, *SIGMOD Record*, 40(4): 21–27.

Mocanu, D., Baronchelli, A., Perra, N., Gonçalves, B., Zhang, Q., and Vespignani, A. (2013) The twitter of Babel: Mapping world languages through microblogging platforms, *PLoS One*, 8(4): e61981. doi:10.1371/journal.pone.0061981.

New York Times (2011) Spotlight again falls on web tools and change, *New York Times*, Available at http://www.nytimes.com/2011/01/30/weekinreview/30shane.html?_r=1&hp (accessed on July 23, 2012).

Nielsen (2012) State of the media: The social media report, Available at http://slidesha.re/Zrwu6V (accessed on June 1, 2013).

Norris, C., McCahill, M., and Wood, D. (2004) The growth of CCTV: A global perspective on the international diffusion of video surveillance in publicly accessible space, *Surveillance & Society*, 2(2/3): 110–135.

Page, L., Brin, S., Motwani, R., and Winograd, T. (1999) The PageRank citation ranking: Bringing order to the web, Technical report. Stanford InfoLab, Stanford, CA.

Quercia, D., Capra, L., and Crowcroft, J. (2012) The social world of twitter: Topics, geography, and emotions, *Proceedings of the Sixth International Conference on Weblogs and Social Media*, Dublin, Ireland, pp. 298–305.

Rahm, E. and Do, H.H. (2000) Data cleaning: Problems and current approaches, *IEEE Data Engineering Bulletin*, 24(4): 3–13.

Raymond, E.S. (1999). *The Cathedral and the Bazaar: Musings on Linux and Open Source by an Accidental Revolutionary*, Cambridge, MA: O'Reilly.

Reinhardt, W., Varlemann, T., Moi, M., and Wilke, A. (2010) Modeling, obtaining and storing data from social media tools with artefact-actor-networks, *Proceedings of ABIS 2010: The 18th International Workshop on Personalization and Recommendation on the Web and Beyond*, Kassel, Germany.

Romero, D., Meeder, B., and Kleinberg, J. (2011) Differences in the mechanics of information diffusion across topics: Idioms, political hashtags, and complex contagion on twitter, *Proceedings of the 20th International Conference on World Wide Web*, Hyderabad, India, pp. 695–704.

Sahito, F., Latif, A., and Slany, W. (2011) Weaving twitter stream into linked data: A proof of concept framework, *Proceedings of the Seventh International Conference on Emerging Technologies*, Islamabad, Pakistan, pp. 1–6.

Sakaki, T., Okazaki, M., and Matsuo, Y. (2010) Earthquake shakes twitter users: Real-time event detection by social sensors, *Proceedings of the 19th International Conference on World Wide Web*, Raleigh, NC, pp. 851–860.

Sapiro, G. (2011) Images everywhere: Looking for models: Technical perspective, *Communications of the ACM*, 54(5): 108–108.

Scott, J. (2012) *Social Network Analysis* (3rd edn.), London, U.K.: SAGE Publications.

Sharma, P., Khurana, U., Shneiderman, B., Scharrenbroich, M., and Locke, J. (2011) Speeding up network layout and centrality measures for social computing goals, in Greenberg, A.M., Kennedy, W.G., and Bos, N.D. (eds.), *Social Computing, Behavioral-Cultural Modeling and Prediction* (*Proceedings of the Sixth International Conference, SBP 2013, Washington, DC, April 2–5, 2013*), Berlin, Germany: Springer-Verlag, pp. 244–251.

Shimojo, A., Kamada, S., Matsumoto, S., and Nakamura, M. (2010) On integrating heterogeneous lifelog services, *Proceedings of the 12th International Conference on Information Integration and Web-based Applications & Services*, New York, pp. 263–272.

Smith, A. (2011) Why Americans use social media: Social networking sites are appealing as a way to maintain contact with close ties and reconnect with old friends, Pew Research Center, Washington, DC, Available at http://bit.ly/rLCsA6 (accessed on June 3, 2013).

Standby Task Force (2011) The security and ethics of live mapping in repressive regimes and hostile environments, Available at http://blog.standbytaskforce.com/?p = 259 (accessed on May 27, 2013).

Stefanidis, A., Cotnoir, A., Croitoru, A., Crooks, A.T., Radzikowski, J., and Rice, M. (2013a) Statehood 2.0: Mapping nation-states and their satellite communities through social media content, *Cartography and Geographic Information Science*, 40(2): 116–129.

Stefanidis, A. and Nittel, S. (2004) *GeoSensor Networks*, Boca Raton, FL: CRC Press.

Stefanidis, T., Crooks, A.T., and Radzikowski, J. (2013b) Harvesting ambient geospatial information from social media feeds, *GeoJournal*, 78(2): 319–338.

Sui, D. and Goodchild, M. (2011) The convergence of GIS and social media: Challenges for GIScience, *International Journal of Geographical Information Science*, 25(11): 1737–1748.

Sun, A., Bhowmick, S.S., Nguyen, N., Tran, K., and Bai, G. (2011a) Tag-based social image retrieval: An empirical evaluation, *Journal of the American Society for Information Science and Technology*, 62(12): 2364–2381.

Sun, Y. and Han, J. (2012) Mining heterogeneous information networks: Principles and methodologies, *Synthesis Lectures on Data Mining and Knowledge Discovery*, 3(2): 1–159.

Sun, Y., Han, J., Yan, X., Yu, P.S., and Wu, T. (2011b) Pathsim: Meta path-based top-k similarity search in heterogeneous information networks, *Proceedings of the 37th Very Large Data Base Conference*, Seattle, WA.

TechAmerica (2012) Demystifying big data: A practical guide to transforming the business of government, *TechAmerica Foundation, Federal Big Data Commission*, Washington, DC, Available at http://bit.ly/11CpPDc (accessed on May 27, 2013).

TechCrunch (2012) How big is Facebook's data? 2.5 billion pieces of content and 500+ terabytes ingested every day, Available at http://tcrn.ch/NhjAVz (accessed on October 29, 2013).

Tobler, W. (1970) A computer movie simulating urban growth in the Detroit region, *Economic Geography*, 46(2): 234–240.

TomTom (2011) This is what we really do with your data, Available at http://www.tomtom.com/page/facts (accessed on June 3, 2013).

Tonkin, T., Pfeiffer, H.D., and Tourte, G. (2012) Twitter, information sharing and the London riots? *Bulletin of the American Society for Information Science and Technology*, 38(2): 49–57.

Tumasjan, A., Sprenger, T., Sandner, P., and Welpe, I. (2011) Election forecasts with twitter: How 140 characters reflect the political landscape, *Social Science Computer Review*, 29(4): 402–418.

Twitter (2011) 200 million tweets per day, Available at http://bit.ly/laY1Jx (accessed on January 19, 2013).

Vatsavai, R.R. and Bhaduri, B. (2013) Geospatial analytics for big spatiotemporal data: Algorithms, applications, and challenges, *NSF Workshop on Big Data and Extreme-Scale Computing*, Charleston, SC.

Wasserman, S. and Faust, K. (1994). *Social Network Analysis: Methods and Applications*, New York, NY: Cambridge University Press.

Wayant, N., Crooks, A.T., Stefanidis, A., Croitoru, A., Radzikowski, J., Stahl, J., and Shine, J. (2012) Spatiotemporal clustering of social media feeds for activity summarization, *GI Science* (*Seventh International Conference for Geographical Information Science*), Columbus, OH.

West, G. (2013) Big data needs a big theory to go with it, *Scientific American*, May 15, Available at http://bit.ly/15wqygI (accessed on May 20, 2013).

Wohn, D.Y., Ellison, N.B., Khan, M.L., Fewins-Bliss, R., and Gray, R. (2013) The role of social media in shaping first-generation high school students' college aspirations: A social capital lens, *Computers & Education*, 63: 424–436.

YouTube (2013) YouTube statistics, Available at http://bit.ly/XQlBYW (accessed on October 29, 2013).

Zhao, S., Zhong, L., Wickramasuriya, J., and Vasudevan, V. (2011) Human as real-time sensors of social and physical events: A case study of Twitter and sports games, Technical report TR0620-2011, Rice University and Motorola Labs, Houston, TX, arXiv:1106.4300v1.

Zook, M., Graham, M., Shelton, T., and Gorman, S. (2010) Volunteered geographic information and crowdsourcing disaster relief: A case study of the Haitian earthquake, *World Medical & Health Policy*, 2(2): 7–33.

12 Insights and Knowledge Discovery from Big Geospatial Data Using TMC-Pattern

Roland Assam and Thomas Seidl

CONTENTS

12.1 INTRODUCTION

The world of mobile computing and communication is evolving in a fast pace. Due to the fact that big geospatial data can mimic behaviors, affiliations, and interests of an individual or a group, there is an enormous research interest to detect mobility patterns or correlations hidden in such data. Modeling trajectories from users' past location histories is of great importance because most location mining tasks such as location prediction, location recommendation, location classification, or outlier detection strongly rely on an underlying trajectory model or scheme. Failing to properly model trajectory data from users' past location histories will have adverse effects in the aforementioned location mining tasks. Data generated by ubiquitous devices are usually called spatiotemporal data. However, this data often consists of more than three dimensions (space and time). That is, mobility data do not only associate location to time, but they also bind location and time to one or more additional attributes (or dimensions). The additional attributes can range from social network activities, Twitter messages, diseases, environmental pollution, age, and telephone calls to SMS messages as depicted in Figure 12.1.

Numerous techniques have been proposed for trajectory modeling (Ashbrook and Starner 2003, Yu et al. 2009) and for mobility data extraction (Giannotti et al. 2007, Nicholson and Noble 2008, Yavas et al. 2005). Most of these state-of-the-art techniques are novel. However, they focus only on the spatiotemporal dimensions to extract or build a trajectory model from raw geospatial data. Social network activities (e.g., Twitter messages, Facebook activities), telephone calls, or disease attributes are

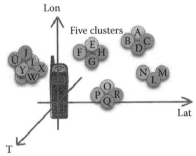

Only spatiotemporal dimensions

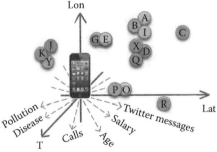

Smart phones: high-dimensional geodata

FIGURE 12.1 Dimensions of state-of-the-art devices.

not considered for trajectory modeling even if they are included in the raw mobility data. Big geospatial data emanating from smart phones and other state-of-the-art ubiquitous devices have a finite but wide variety of dimensions. Since existing trajectory models (Ashbrook and Starner 2003, Giannotti et al. 2007, Yu et al. 2009) are unable to handle or utilize these non-spatiotemporal dimensions, we strongly believe a new trajectory model that incorporates and processes dimensions beyond space and time is required to meet the demands of this rich diverse multidimensional big geospatial data. Due to this observation, we propose a novel trajectory model called Time Mobility Context Correlation Pattern (TMC-Pattern).

TMC-Pattern considers a wide variety of dimensions and utilizes subspace clustering (SUBCLU) to find contextual regions of interest (ROI). In addition, TMC-Pattern rigorously captures and embeds infrastructural, human, social, and behavioral patterns hidden in big geospatial data into the trajectory model. Furthermore, TMC-Pattern incorporates the context of a user and a region into the pattern.

Table 12.1 shows the differences between our proposed TMC-Pattern and some state-of-the-art trajectory models. There are three major observations in Table 12.1. First, TMC-Pattern utilizes SUBCLU, while the other models use full space clustering. The benefits of employing SUBCLU will be elaborated later. Secondly, it incorporates network interconnectivity between regions via the use of the betweenness centrality measure. Thirdly, it uses new notions such as stay mode (SM), pass mode (PM), and brief stop mode (BSM) to portray the state of an object in a region.

Furthermore, in many existing trajectory models (Ashbrook and Starner 2003, Giannotti et al. 2007, Quannan et al. 2008, Yu et al. 2009), the significance or

TABLE 12.1
Comparison of Trajectory Models

	TMC-Pattern	T-Pattern (Giannotti et al. 2007)	Li et al. (Quannan et al. 2008)	Ashbrook et al. (Ashbrook and Starner 2003)
Dimensions	>3	3	3	3
Dimension type	Space, time, social network activity, disease, SMS message, phone call	Space and time	Space and time	Space and time
Significant regions	SM, PM, BSM	Time interval annotation	Space and time	Space and time
Embed infrastructure	Yes	No	No	No
Frequent regions	PrefixSpan	PrefixSpan and MISTA	Hub based	—
Clustering	SUBCLU	Grid and DBSCAN (full space clustering)	Hierarchal clustering (full space clustering)	K-means (full space clustering)

importance of a region is determined by the stay *time*. The stay time refers to the duration that an object stays in a region. TMC-Pattern is proposed due to the *over-generalization* of mobility information on existing trajectory models. For example, in the trajectory model proposed by Quannan et al. (2008), a region is termed a stay region if the duration of an object in that region exceeds a defined threshold time (e.g., 20 min). People usually spend longer than 20 min in a theater. Hence, a theater will be labeled according to their model as an important or interesting stay region. But is a theater truly a stay region throughout the day? Theaters are almost empty in the morning and are barely full during the afternoon. Mining techniques for big geospatial data must provide precise insights to enable accurate decision making. Quannan et al. (2008) interpretation is good for interesting location mining but is very vague to predict the next location of a user. A better depiction would be to include some time-related information to the location. For instance, *theaters are interesting location "as from evening."* TMC-Pattern correlates time periods to significant or important regions. Ashbrook and Starner (2003) and Nicholson and Noble (2008) use route frequencies to derive the Markov model's probabilities to predict the next location of an object. Looking at the overall frequency by which a route is taken, one can quickly conclude that because of its popularity, it is the best route. However, the traversals in the morning and afternoon might account for the high route frequency, while other routes might have higher traversal frequencies at night, due to social or human behaviors such as crime risk or better night view of the city. Predictions based on highest frequency without the period of the day can very often be misleading and decrease prediction accuracy.

TMC-Pattern can be employed to create value from big geospatial data in a range of far-flung fields such as mobile social networking, location-based advertisement, telecommunication, healthcare patient monitoring, traffic monitoring, and trend analysis. Besides, unlike most existing trajectory modeling techniques, TMC-Pattern can be used to study the relationship between pollution and mobility and how mobility impacts the spread of disease.

In this chapter, we provide a novel technique to model big geospatial data and two techniques to mine and extract knowledge and insights from big geospatial data. Modeling raw and diverse big location-related data is quite challenging in terms of quality and runtime efficiency. Existing techniques are forced to choose a certain level of granularity for trajectory modeling by the use of threshold stay time to demarcate important from less important locations. With respect to the granularity of modeling raw data, most models (Ashbrook and Starner 2003, Quannan et al. 2008) are focused on the *big picture*, that is, they discard numerous geospatial points to obtain the overview of an object's mobility. In this big data era, it is difficult, yet important, to grasp or at least know which object's movement can be game changing (e.g., for a location-based advertisement or recommendation system) during decision making. This underscores the need for TMC-Pattern that captures even a slight movement of an object and all its nearest neighbors, as well as the big picture of mobility.

On the other hand, to mine and uncover patterns from big geospatial data, we propose an approach that employs TMC-Pattern and PrefixSpan (Pei et al. 2001) to discover context-aware ROI. We also introduce a novel technique that uses TMC-Pattern and Markov model to predict future locations. Furthermore, in this chapter, we demonstrate, using a real dataset, how Twitter messages can be used to track the existence or spread of a disease. Specifically, we utilize tweets from Twitter that consist of the *flu* keyword and originate from a defined geographic location and then find novel correlations between location and the possible spread of the flu disease.

Apart from TMC-Pattern, a wide range of techniques and methodologies has been proposed for geospatial data modeling and knowledge discovery through data mining. T-Pattern (Giannotti et al. 2007) is a novel trajectory pattern modeling technique that partitions a geospatial data into different regions based on frequently visited sequence and a travel time. Han and Cho (2006) proposed a trajectory model that uses multiple users' location dataset to build movement pattern. In terms of context extraction, Eagle and Pentland (2006) demonstrate how context can be inferred from a user's daily movement and from socially significant locations. Other works that focus on semantic location extraction and context inference include Kalasapur et al. (2009) and Liao et al. (2007). Our proposed TMC-Pattern is quite different from the movement pattern of Giannotti et al. (2007) and Han and Cho (2006) in that it strongly correlates mobility to time periods and uses object modes and network properties to instill realistic human, social, and infrastructural information about a user and the current location or region in the trajectory model. Quannan et al. (2008) and Yu et al. (2009) proposed a trajectory modeling technique that models the location history of a user and finds similarities between multiple users. They utilized the notion of stay point to create geo-regions where objects stay for a given period of time and then employ a hierarchy clustering algorithm to group multiple objects in these regions. While their model uses the notion of stay time to get stay regions, our model extends the notion of stay time and proposes novel notions like PM and brief stay mode to capture passersby. Besides, we use SUBCLU to project the activity of a region from different perspectives. Other works that propose an approach for extracting location and building statistical trajectory models include Minkyong et al. (2006) and Laasonen et al. (2004). Monreale et al. (2009) proposed a novel technique that uses the global trajectory history of all users to predict the next location of an object. Unlike Monreale et al. (2009), our technique utilizes a second-order Markov chain for location prediction. The predictive models of Ashbrook and Starner (2003), Jungkeun et al. (2006), and Asahara et al. (2011) are based on Markov model, and our technique differs from them in that we use a different score metric to compute the transition probabilities. Yavas et al. (2005) and Morzy (2006) proposed methods to predict future locations using association mining rules.

12.2   TRAJECTORY MODELING

Our primary objective for modeling trajectory of big geospatial data is to extract and model the trajectory history of an object to capture the mobility pattern of the object and its frequent nearest neighbors.

12.2.1   TMC-Pattern

In this section, we present TMC-Pattern, a novel trajectory modeling technique. Figure 12.2 depicts an overview of this data extraction and modeling technique.

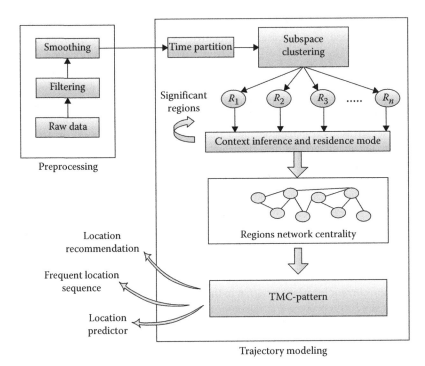

FIGURE 12.2   Geospatial data extraction and modeling using TMC-Pattern.

12.2.1.1   Determining Meaningful Location

In many previous works, for example, Ashbrook and Starner (2003), Yu et al. (2009), and Jungkeun et al. (2006), significant or interesting locations are extracted from large mobility datasets by selecting locations where an object stays for at least a given period of time (usually termed *stay time*). Stay time is the time interval between two successive traces of an object's trajectory. TMC-Pattern also utilizes stay time to identify *meaningful* locations from raw mobility data, however, in a different way. In TMC-Pattern, if an object resides for more than a threshold stay time in a given region, the object is said to be at a SM in that region.

Definition 12.1

SM is the mode of an object in a given region with its duration of stay in that region is greater or equal to the threshold stay time T_{st}.

In previous works, the notion of stay time is overgeneralized. They discard or ignore traces if an object's stay time in a region is short. For example, Figure 12.3 illustrates the trajectory of a person that visits six locations and the time spent at each location. From Figure 12.3, a trajectory model that considers a location interesting

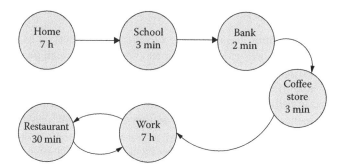

FIGURE 12.3   A journey from home to work.

or significant only if the stay time threshold is at least 5 min will have a trajectory sequence *Home ↦ Work ↦ Restaurant*, which fails to capture the person's short visits at the bank, school, and coffee shop, simply because their stay times are less than 5 min. We stress that this example reflects real human mobility, as visiting an ATM machine at the bank or dropping a kid at school or picking up a coffee in most cases takes less than 5 min.

Liao et al. (2007) alluded that points could be considered as stay points due to traffic congestion. Intuitively, if the threshold stay time is less than 5 min, the rate of such misleading stay point information becomes higher. However, we argue that failing to integrate such important locations (under 5 min) during data extraction truly results in information loss, which can, among other things, degrade location prediction accuracy. To address this information leakage, we propose a new threshold time, namely, threshold pass time (T_{pt}), as well as two new modes called BSM and PM.

Definition 12.2

PM is the mode of an object in a given region with its duration of stay in that region is less than the threshold pass time (T_{pt}).

Definition 12.3

BSM is the mode of an object in a given region with its duration of stay in that region is greater than the threshold pass time (T_{pt}) but less than the threshold stay time (T_{st}).

Throughout this chapter, the term *residence mode* will be used to refer to SM, PM, and BSM. In fact, our main rationale for these residence modes is to capture *flash of movements* exhibited by neighboring objects (e.g., passersby) that traverse certain regions with or without a short stop. Acquiring such neighboring objects is vital and pivotal in the determination of frequent neighbors.

12.2.2 Time Correlation

In Section 12.1, we mentioned that labeling a theater as an important location can be misleading to a location predictor. We also mentioned that a route might be very popular but least used at night because of crime or other social or infrastructural factors. Failing to correlate time to regions or infrastructures ignores and does not reflect human behavioral pattern in a trajectory model. This may decrease the accuracy of location recommendation and location prediction systems. TMC-Pattern correlates time to regions and infrastructures by associating a time period to each residence mode (SM, BSM, PM) in a given region. Specifically, we segment time into a consistent and finite set of time periods or intervals. For example, a day can be partitioned into 24 h, or a month can be segmented into days or weeks.

Definition 12.4

Time period is a finite set T that consists of time intervals $t_1, t_2, t_3, ..., t_n$, and the time intervals are formed by breaking up the original time (T_ϕ) into several time ranges.

The time intervals of the different elements of a time period must not be the same, and each time interval can be represented by a label. For example, in this work, we use a TMC-Pattern whose time period is given by the set $T = \{M, A, E, N\}$ where the time labels M, A, E, N refer to morning, afternoon, evening, and night, respectively. Set T is created by the segmentation of T_φ. Morning (M) is the time interval from 7:01 a.m. to 11:59 a.m., while afternoon (A) is the time range from 12:00 p.m. to 5:59 p.m. Likewise, evening (E) corresponds to the time frame from 6:00 p.m. to 11:59 p.m., and finally, night (N) is considered as the time interval from 12:00 a.m. to 7:00 a.m. The partitioning of T_ϕ into different time ranges during the derivation of time period should depend on the data mining task. As explained earlier, the set T for our purpose uses only four time ranges.

12.2.3 Location Context Awareness

TMC-Pattern entails the context of a user or a region. The context of a user in a given region strongly depends on time. This is another rationale why TMC-Pattern correlates mobility to time and strives to find the residence mode of a user in a given region. The context of a user refers to the activities or the state of mind of the user at a given time and region. Likewise, the context of a region refers to a finite set of events that occur in the region (at a given time). A region is usually visited by many people and depending on the size of the region, the context of the region might differ from one individual to another.

Definition 12.5

TMC context is a finite set C_{tmc} of activities or events or information $i_1, i_2, i_3 \cdots i_m$ about a region and the user at a given time period T.

Context information about a region can be extracted using the IP or GPS position of the region. For example, the context of a region consisting of a mall is given by $C_{tmc} = \{Shopping, Restaurant\}$. Context information about a user could be inferred from SMS messages, Twitter messages, social network activities, telephone calls, etc. If no activity or information about a region or a user in that region could be inferred during preprocessing, then $C_{tmc} = \varnothing$.

12.2.4 RELEVANCE MEASURES OF A REGION

Regions are hosts to objects and at times their nearest neighbors. To determine the importance of a region, and to strongly infuse the infrastructural properties of a region into TMC-Pattern, we borrow principles of graph and network theory. A region can be modeled as a *node* and the transition from one region to another as an *edge*. In network theory, there are several measures that can be used to determine the relevance of a node (region). These include shortest path, eigenvalue, node degree, closeness, and betweenness. According to research studies conducted by Kitsak et al. (2010), the betweenness is the most important network measure. This work showed that even though a path can be the shortest path, the likelihood of utilizing the path is more dependent on its betweenness measure than the shortest path measure. Such network connectivity principles are immensely valuable for a location prediction technique. Towards this end, TMC-Pattern utilizes the betweenness to depict the importance of a region. The betweenness centrality measure is based on the shortest path. The betweenness of node b is given by

$$f(b) = \sum \frac{\sigma_a \mapsto c(b)}{\sigma_a \mapsto c}$$

where $\sigma_a \mapsto c$ is the sum of all shortest paths from node a to node c, while $\sigma_a \mapsto c(b)$ is the number of the latter shortest paths that pass through b.

The other relevance measures adopted by TMC-Pattern are user frequency and node frequency. User frequency refers to the frequency at which a user has visited or passed through a region for *a particular time period*, whereas region frequency corresponds to the total frequency by which all objects have visited or passed through a region *for a given time period* (e.g., 11/770 denotes that a user frequency in a region is 11, while the region frequency is 770). As a summary, TMC-Pattern uses betweenness, region frequency, and user frequency as its relevance measures. These relevance measures reveal the importance of a region or the transition between two regions and are used in Section 12.3.2 to assign weights to nodes and edges for prediction.

12.2.5 TMC-PATTERN

In this section, we present TMC-Pattern.

Definition 12.6

Time Mobility Context Correlation Feature (TMC-Feature) *is a 7D vector summarizing mobility information about an object in a given region and it is represented by*

$$\langle R, \quad u, \quad \textbf{\textit{TP}}, \quad \textbf{\textit{RM}}, \quad \textbf{\textit{UNF}}, \quad \textbf{\textit{C}}, \quad \textbf{\textit{NB}} \rangle$$

where
  R is a region ID or label
  u is the user ID
  TP *denotes the time period*
  RM *is the residence mode*
  UNF *refers to the user and node frequency*
  C *and* **NB** *stand for TMC context in Definition 12.5 and node betweenness, respectively*

Example: Consider that the time period $T = \{M, A, E, N\}$ is used as indicated in Section 12.2.2. Let the object ID of an object O be 10 and the region ID of region A be RegA and its betweenness be 0.7, given that object O resides in region A with the residence mode SM in the morning (i.e., $TP = M$) and its TMC context is $C_{tmc} = \{Bar, Shopping\}$. If object O's frequency at region A during this time period is 114, and the node frequency of region A is 1400, then the TMC-Feature of O is

$$\langle RegA, 10, M, SM, 114/1400, \{Bar, Shopping\}, 0.7 \rangle$$

A TMC-Feature reflects the human, social, and infrastructural characteristics of an object in a region. A TMC-Feature provides effective and sufficient information needed to make prediction or recommendation decisions. Hence, our location prediction algorithm utilizes it, and this greatly alleviates the burden from our prediction algorithm to rigorously mine raw trajectory data.

Definition 12.7

TMC-Pattern *is a sequence of two or more TMC-Features that correspond to the trajectory of an object. It is represented by*

$$\langle R_1, u, \textbf{\textit{TP}}, \textbf{\textit{RM}}, \textbf{\textit{UNF}}, \textbf{\textit{C}}, \textbf{\textit{NB}} \rangle \mapsto \langle R_2, u, \textbf{\textit{TP}}, \textbf{\textit{RM}}, \textbf{\textit{UNF}}, \textbf{\textit{C}}, \textbf{\textit{NB}} \rangle$$

$$\ldots \mapsto \langle R_1, u, \textbf{\textit{TP}}, \textbf{\textit{RM}}, \textbf{\textit{UNF}}, \textbf{\textit{C}}, \textbf{\textit{NB}} \rangle$$

where R_i is the region ID or label and **TP, RM, UNF, C,** *and* **NB** *are w.r.t. the region ID R_i as in Definition 12.6.*

12.2.5.1   Determining Residence Mode of a Region

Estimating the correct residence mode of an object in a region at a given time period is very crucial in predicting the next location. In the last section, we presented the TMC-Pattern of an object in a given region. It is easy to retrieve the residence mode from a single TMC-Feature. However, intuitively, if an object transverses a region several times (different days), it is possible that at times, the object stays in a region (SM), makes a brief stop in the region (BSM), or passes quickly through it (PM). On several occasions, determining the residence mode when an object makes multiple visits to a given region with different residence modes can be ambiguous. For example, assume object O visits region A frequently with the following TMC-Features from Monday to Friday, respectively:

$$\langle RegA, 10, M, SM, 114/1400, \{Work, Bar\}, 0.7 \rangle$$

$$\langle RegA, 10, M, BSM, 115/1470, \{Work, Bar\}, 0.7 \rangle$$

$$\langle RegA, 10, M, BSM, 116/1550, \varnothing, 0.7 \rangle$$

$$\langle RegA, 10, M, BSM, 117/1571, \{Work, Bar\}, 0.7 \rangle$$

$$\langle RegA, 10, M, SM, 118/1590, \{Bar, Shopping\}, 0.9 \rangle$$

From these TMC-Features, the residence mode of object O in the morning in region A varies; hence, the residence mode is ambiguous. As a result, a net residence mode of object O in region A has to be determined. We tackle this ambiguity by utilizing the association mining notion of confidence.

Definition 12.8

Residence confidence ($Conf_{RM}$) of a region at a given time period from a list of TMC-Features is given by

$$Conf_{RM} = \frac{\text{Count of specific residence mode}}{\text{Total count of residence modes}} \qquad (12.1)$$

The net residence mode of a region is determined using Equation 12.1 and a predefined minimum confidence ($minConf$) as follows. First, we compute $Conf_{RM}$ for each residence mode (i.e., SM, BSM, or PM) in the list of TMC-Features, and we then select the residence mode with the highest confidence ($Conf_{RM}$). Next, we determine if the latter confidence (i.e., $maxConf_{RM}$) is higher or the same as the $minConf$. If it is higher (i.e., $maxConf_{RM} \geq minConf$), then the residence mode with that confidence is considered as the net residence mode of the region.

Example 1: Given the $minConf$ of 60%, using the earlier TMC-Feature list, the net residence mode of object O in region A is computed as follows:

$$SM\ Conf_{RM} = \frac{\text{Count of } SM}{\text{Total residence modes}} = \frac{2}{5}$$

$$BSM\ Conf_{RM} = \frac{\text{Count of } BSM}{\text{Total residence modes}} = \frac{3}{5}$$

Since BSM has the highest confidence and $BSM\ Conf_{RM} \geq minConf$, it means the net residence mode for region A is BSM.

However, if the highest confidence from Equation 12.1 is lower than $minConf$, we use certain *precedence rules* to address this ambiguity based on the following intuition. BSM is an intermediate between PM and SM. If an object is captured in the SM for a surmountable amount of time in a region, even if the object is in the PM (or BSM) for almost that period of time, from the human behavior perspective, it is logical to represent the net residence mode of the object in that region as SM.

On the other hand, if the number of PMs is comparable to the BSMs, the residence mode is considered as a PM because of the object's tendency to spend less time in that region. Hence, SM $\preccurlyeq$ PM $\preccurlyeq$ BSM. That is, SM precedes PM that in turn precedes BSM. The precedence rules are as follows:

1. SM $\vee$ BSM $\vee$ PM $\Rightarrow$ SM
2. SM $\vee$ BSM $\Rightarrow$ SM
3. BSM $\vee$ PM $\Rightarrow$ PM

Example 2: Using the same assumptions in Example 1 except that the $minConf$ is now 70%, since the highest confidence will now be less than $minConf$, we utilize the aforementioned precedence rules to determine the net residence mode. Based on these rules, the net residence mode of object O in region A is SM. For the next location prediction, inferring the correct residence mode of object O in region A can serve as an important hint to predict correctly.

12.2.6 TRAJECTORY EXTRACTION

TMC-Pattern is extracted and modeled from raw spatiotemporal data as shown in Figure 12.2 by using Algorithm 12.1. Given a mobility dataset that comprises multiuser trajectories, the dataset is first subdivided into single-user data (Line 2–Line 9). We determine the residence mode of each raw point.

Algorithm 12.1 TMC Data Extraction and Modeling

Input: Trajectory dataset D, time interval T_s, eps-radius $\in$, minimum point
MinPts, relevant dimensions *dim*, minimum residence confidence
minConf

Output: Cluster $C = \{c_1, c_2, \ldots, c_n\}$ //Each c_i is an arbitrary-shaped region

1. $D_{multi} \leftarrow \varnothing$;   //Multiuser data buffer D_{multi}
2.   **for all** (D_u) **do**   //Processing Single User Dataset D_u
3.   Consider single user dataset D_u from D;
4.   Determine Residence Mode of raw points in D_u;   //Section 12.2.5.1
5.   Discard PM points between T_s in D_u;
6.   Get moving average of raw points in D_u;
7.   Partition *Time* attribute in D_u into $TP = \{M, A, E, N\}$;   //Section 12.2.2
8. $D_{multi} = D_{multi} \bigcup D_u$;
9.   **end for**
10. $C \leftarrow getTMCClusters(dim, D_{multi}, \in, MinPts)$;
11.   **for all** $(c_i \subset C)$ **do**
12. $listOfObjects \leftarrow getListOfObjects(c_i)$;
13.   **for all** (*objId* in *listOfObjects*) **do**
14. $objTS \leftarrow getTimePeriod(objId)$;
15. $tmcContext \leftarrow getContext(objId, c_i)$;   //Section 12.2.3
16. $regionNetwork \leftarrow getNetwork(objId, c_i, C)$;   //Section 12.2.4
17.   **end for**
18.   **end for**
19.   **return** Cluster $C = \{c_1, c_2, \ldots, c_n\}$

Logically, reading the geographical positions of an object after a very short interval like 2, 5, or 10 s cannot profoundly contribute to the big picture of an object's movement. Thus, discarding such raw points leads to negligible information loss. For this, we filter raw data by choosing certain points from the list of points after a time interval T_s and discard points with T_s whose *residence mode* = PM (Line 4–Line 5). T_s can be 10, 20, or 30 s, depending on the number of points in the dataset or sampling rate. Since the measurements of GPS positions and Wi-Fi locations are associated with errors, we utilize the moving average filter to smoothen or estimate a more precise location of the mobility data (Line 6). After more accurate geographical positions of the traces have been determined, we categorize the time attribute of each trace to a given time period TP where $TP = \{M, A, E, N\}$ (Line 7).

The same procedure is repeated to filter, estimate, and categorize the traces of all other users, and after each iteration, the filtered and categorized data of a single user are added to the multiuser dataset buffer D_{multi} (Line 8). After the raw trajectory dataset of all users have been preprocessed to build D_{multi}, we use the SUBCLU (Kailing et al. 2004) algorithm to find clusters of the points for different time periods. This is accomplished by inputting D_{multi} to the SUBCLU algorithm together with the relevant dimension (Line 10). This results in the formation of one or more arbitrary-shaped clusters. Each cluster corresponds to a region (also called TMC-Region).

We compute the TMC context (Line 15) and network properties (Line 16) of each object at each TMC-Region at a given time period. The final result of Algorithm 12.1 is the generation of one or more clusters (Line 19).

As mentioned earlier, mobility data comprise spatiotemporal dimensions and other dimensions such as phone calls, social network activities, Twitter messages, and diseases. With the use of SUBCLU, we can select relevant dimensions and depict a TMC-Pattern based on the subspaces instead of the full space. For example, we might be interested only in a region or location visited by people under 30, or we are interested to find the correlation between a location and a social network activity. In scenarios where mobility data have more than three dimensions, the ability to select only interesting dimensions (instead of the full space) gives TMC-Pattern trajectory model a higher advantage over existing trajectory models of, for example, Giannotti et al. (2007), Yu et al. (2009), and Ashbrook and Starner (2003) that use full space.

12.3 TRAJECTORY MINING

12.3.1 FREQUENT LOCATIONS FROM TMC-PATTERN

Frequent location subsequences, or ROI, can be discovered using sequential pattern mining. There are several techniques to mine sequential patterns. Due to the fact that location datasets are inherently large (thus Big Data), we adopt the PrefixSpan (Pei et al. 2001) pattern mining technique, since it does not involve candidate generations and has the best overall performance. Given a minimum support, frequently occurring regions can be extracted from TMC-Pattern.

Algorithm 12.2 FrequentTMCPattern $(\alpha, l, D_\alpha, Sup\_min, C)$

> **Input:** Subspace clustered regions C, α-projected database, pattern length l, minimum support Sup_min
>
> **Output:** Sequence of ROI $R = r_1 \rightarrow r_2 \rightarrow \dots \rightarrow r_n$

1. $R \leftarrow \emptyset; D_{seq} \leftarrow \emptyset;$
2.   **if** $(C! = NULL)$ **then**   /* Sequence Database D_{seq} is created once */
3. $D_{seq} \leftarrow getRegionSequences(C);$
4.   **end if**
5. $F_{reg} \leftarrow \{\textbf{\textit{frequent Regions}}$ in $D_{seq}\};$
6.   **for all** $(r_i \in F_{reg})$ **do**
7. $\alpha = (\beta_1 \rightarrow \dots \beta_n \cup \{r_i\});$
8.   **if** $(Support(\alpha, D_{seq}) \geq Sup\_min)$ **then**
9. $F_{reg} \leftarrow F_{reg} \bigcup \alpha;$
10. $D|_\alpha \leftarrow \alpha\text{-}projected\ Database;$
11. $R = R \bigcup frequentTMCPattern(\alpha, l+1, D|_\alpha, Sup\_min, NULL);$ // PrefixSpan Recursion
12.   **end if**
13.   **end for**
14.   **return R**

Algorithm 12.2 depicts a customized version of the PrefixSpan algorithm that is utilized to find frequent TMC-Regions. While the original PrefixSpan algorithm deals with frequent items, our customized version aims at discovering frequent TMC-Region sequences. The first step of Algorithm 12.2 is to create a mobility sequence database (Line 3). Subspace clusters that correspond to TMC-Regions are used to form a sequence database. This is accomplished by representing each trip made by a user as a sequence of TMC-Regions. Each TMC-Region in the sequence database is chronologically listed w.r.t. the time and direction of motion. For example, assume a trip T1 where a user passes through three TMC-Regions, namely, home (H), school (S), and bank (B) at time t_1, t_2, t_3, respectively, where $t_1 < t_2 < t_3$. Then the region sequence alphabet in the sequence database should be labeled as H → S → B not H → B → S or B → H → S. We should note that TMC-Regions are not trivial regions. Their creation is based on a given time period and they might be associated to one or more context. To mine frequent location subsequences, we employ the prefix and suffix approach of PrefixSpan to construct the projected databases that correspond to a region sequence.

After the sequence database has been created, we start by scanning all 1-length sequence of the sequence database (Line 6–Line 7). If the support of the 1-length sequence (or TMC-Region) is not less than the minimum support (Line 9), the sequence is considered as a prefix. After all 1-length sequences have been determined, the search space is partitioned into m disjoint subsets, where m is the number of 1-length prefix TMC-Regions.

A projected database of a 1-length prefix is constructed (Line 10) for each of the latter disjoint subsets and their sequential patterns are computed. Line 6 runs recursively until all frequent sequential patterns are found. That is, 2-length sequences are found after scanning the 1-length projected databases. All the 2-length sequences are recursively utilized to form their corresponding projected database (Line 11). Scanning a 2-length sequence projected database leads to the creation of 3-length sequence, which is recursively used to form a projected database. The process continues recursively until no more projected database can be formed.

We should note that unlike previous works (Giannotti et al. 2007, Quannan et al. 2008, Yu et al. 2009) where such frequent regions correspond to the popularity of a place (since they considered only spatiotemporal dimensions), frequent TMC-Regions strongly depend on the chosen subspace dimension. For example, a frequent TMC can be a sequence of regions associated to a disease, age, or social network activity.

12.3.2 TMC-Pattern and Markov Chain for Prediction

12.3.2.1 Markov Chains

Markov Chain is a probabilistic model that comprises a finite number of states that are linked to one another as depicted in Figure 12.4a. A transition from state x to state y is associated with a probability called the transition probability, which is represented by T_{xy}.

The transition probability of a Markov chain depends on the order of the Markov chain. For example, a first-order Markov chain depends solely on the present (current)

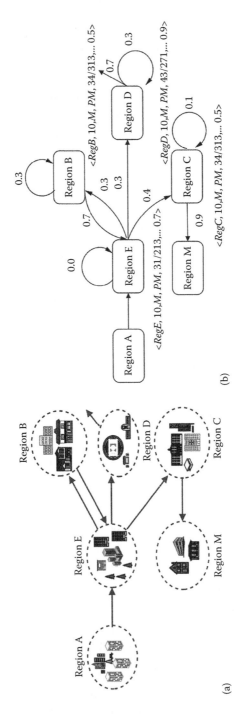

FIGURE 12.4　Probabilistic model from TMC-Pattern. (a) Subspace clustered TMC regions. (b) Markov chain.

state and no past state, while the second-order Markov chain depends on both the current state and the previous state that precedes the current state. In this work, a discrete time second-order Markov chain is utilized. Markov chains are widely used in areas such as bioinformatics and speech recognition.

12.3.2.2 Markov Chain from TMC-Pattern

Figure 12.4b shows the movement of a user from Region A (start) to the current Region E. All regions in Figure 12.4b correspond to the subspace clusters formed during TMC-Pattern trajectory modeling. Using TMC-Pattern, we build a probabilistic model of a user's movement as shown in Figure 12.4b. Based on our second-order Markov chain and the intuition behind it, it means that even though the object's movement started from Region A (r_a), the joint probability distribution required to predict that the object's next location is Region C (r_c), w.r.t. Figure 12.4b, is given by: ($r_c \mid r_a, r_e$).

Real-life location sequences are longer than the example provided in Figure 12.4b. Generally, for a location sequence that starts from Region r_1 to r_k, the joint probability distribution is given by

$$P(r_1 \ldots r_k) = P(r_1)P(r_2|r_1)P(r_3|r_2,r_1)\ldots P(r_n \mid r_{k-1},r_{k-2})$$

In the current region (Region E), the user has the possibility to transit to Region B, Region D, or Region C. The TMC-Pattern of the user's movement is transformed into a Markov chain as depicted in Figure 12.4b. Each state of the Markov chain is represented by its corresponding region in the TMC-Pattern, and the regions are tagged with their respective TMC-Features. To predict the next location, we utilize scores from TMC-Pattern and transition probabilities from pure second-order Markov chain.

12.3.2.3 Computation of Markov Chain Transition Probability

We compute the transition probabilities of a second-order Markov chain by considering the frequency at which the user has visited a region from the current region and the region that precedes the current region. That is, we find the frequency of the sequences $A \mapsto E \mapsto B, A \mapsto E \mapsto D, A \mapsto E \mapsto C$ from the user's mobility history and then compute the transition probabilities based on these frequencies. We should note that the sum of all transition probabilities is 1.

12.3.2.4 Computation of Scores from TMC-Pattern

Let η denote the number of candidate next locations. If the current location is Region E, ($\eta = 3$), the scores of the candidate next regions are computed as follows:

1. The user's TMC-Features history at all candidates' next locations (i.e., Region B, Region D, and Region C) is retrieved. Intuitively, time can play an important role in determining a user's next location. For this, when using the past location history of a user, only locations that have a similar time period to that of the current location will be considered. Hence, the retrieved TMC-Features of the candidate next locations that do not have the same *time period* as that of the current region's TMC-Feature are discarded.

2. We compare the vectors (i.e., user frequency, node frequency, node betweenness) embedded in the TMC-Feature of the current region's TMC-Feature and that of each candidate next location, and scores are awarded. These vectors and their corresponding scores are awarded as follows. The residence mode of the current region is compared to that of each candidate next locations; a match of the residence mode is given a score of $1/2\eta$, while a mismatch is assigned a 0. Scores arising from user frequency, node frequency, and node betweenness are determined by comparing the candidate next locations to each other. The region with the highest user frequency is given a score of $1/\eta$, and all other regions are levied with a score of 0. Likewise, the node frequency and node betweenness of all candidate next locations are compared. For each of these measures, the region with the highest value is given a score of $1/2\eta$, and the others are assigned a score of 0. We sum all the scores derived from the TMC-Features and the transition probability from a second-order Markov chain. The candidate regions with the highest value are considered as the next location.

12.4   EMPIRICAL EVALUATIONS

Experiments were conducted on an Intel 8 GB RAM PC and the algorithms were implemented in Java. To evaluate our TMC-Pattern trajectory model, we based our experimental analysis on three major aspects: (1) extracting and modeling raw mobility data, (2) detecting frequent location (sub) sequences, and (3) location prediction using TMC-Pattern. In each of these aspects, we compare our technique with state-of-the-art works. They include the T-Pattern algorithm (Giannotti et al. 2007), the location extraction and prediction techniques proposed by Ashbrook and Starner (2003), and the trajectory modeling approach proposed by Yu et al. (2009). Throughout this section, we refer to these previous works as TP, UGLSL, and MCBL, respectively.

12.4.1   EXPERIMENTAL DATASET

We conducted our experiments on the GeoLife (Yu et al. 2009) Microsoft Asia human mobility real datasets. The GeoLife dataset entails the mobility history of 165 users mostly around Beijing and China. We utilized the mobility data of 52 users, which comprise of more than 700 K raw GPS points. Furthermore, in order to evaluate the capability of TMC-Pattern for finding contextual ROI (other than location) in social networks, we created a real dataset called Twitter–GeoLife Disease dataset. To create the latter dataset, we utilized longitudes and latitudes from points in the GeoLife dataset to search for the *flu* keyword on the Twitter search API. Our search queries retrieved tweets in year 2009 that were within a radius of 150 km from the inputted coordinates. The dataset consisted of 11 users and 14,247 traces.

12.4.2   EVALUATION OF TMC-PATTERN EXTRACTION

We evaluated the strength and capabilities of our technique to extract and model both single and multiple-user mobility histories from the GeoLife dataset.

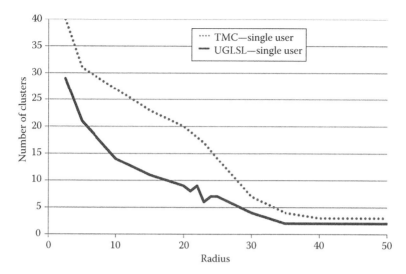

FIGURE 12.5   Single-user extraction.

12.4.2.1   Single-User Data

We conducted several experiments on the trajectory of a single GeoLife user's mobility history. Raw single-user trajectory GPS points are inputted into our trajectory extraction algorithm (Algorithm 12.1). During the extraction process, we avoid huge amounts of redundant trajectory traces by setting $T_s = 2$ min. Since the GeoLife dataset has a sampling rate of 2 s, after the residence mode of each point in the single-user GeoLife dataset has been determined, as described in Section 12.2.6, points tagged as PM within each T_s intervals are discarded. Since the GeoLife dataset consists of only spatiotemporal domains, to find significant regions or TMC-Regions, we selected the space and time domains as the relevant dimensions. We carried out the experiments several times with $2.5 \leq R \leq 50$ km. Figure 12.5 depicts the number of TMC-Regions found. The latter figure shows that as the radius increases, the number of clusters reduces as expected by intuition. It also illustrates the number of clusters found for the same dataset and radii when the customized k-means clustering algorithm of UGLSL (Ashbrook and Starner 2003) is utilized. It shows that our technique detects more clusters than UGLSL for a given radius.

12.4.2.2   Multiuser Data

As mentioned in Section 12.2.6, multiple-user trajectories are extracted and modeled by first preprocessing single-user data. Then, all single-user datasets are combined and clustered using SUBCLU. We processed 52 different user trajectory histories. We set $T_s = 2$ min and utilized an eps-radius (distance of the SUBCLU algorithm) that corresponds to 10 km. We ran the experiments several times for different threshold stay times ranging from 2.5 to 30 min. We performed the same experiments with the same threshold stay times for two state-of-the-art works, MCBL and UGLSL. MCBL and UGLSL employ hierarchy clustering and k-means clustering, respectively. Figure 12.6 depicts the result of this comparison. It shows that

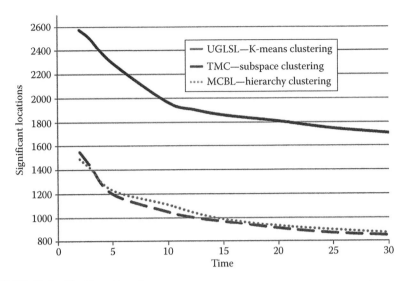

FIGURE 12.6   Significant locations.

the TMC-Pattern extraction technique outperforms MCBL and UGLSL in terms of number of significant locations found for a given threshold time. This can be explained by the fact that during extraction, our approach captures even a flash of movement of an object and its nearest neighbors, while the other techniques ignore points prior to the modeling.

We performed another set of experiments using multiuser data where time period was selected as a relevant dimension. This translates into the formation of clusters such that points are clustered based on the time range ($T = \{M, A, E, N\}$). Figure 12.7 illustrates the result, which shows that more clusters are found in the afternoon than in the morning and evening.

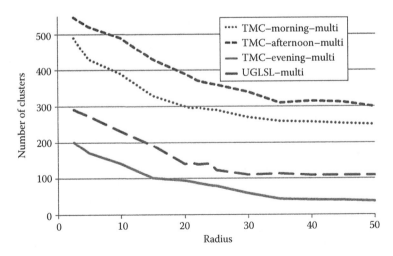

FIGURE 12.7   Clusters versus radius in km.

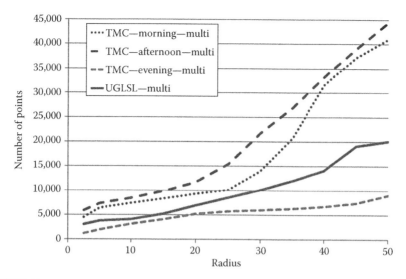

FIGURE 12.8 Number of points versus radius.

MCBL and UGLSL do not have the capability to accomplish this. Furthermore, we analyzed the number of points in a TMC-Region or cluster as depicted in Figure 12.8, which shows that as the radius increases, the number of points in a TMC-Region increases. We compared this with UGLSL and observed that the TMC-Regions had more GPS points for a given radius than UGLSL clusters.

Figure 12.9 depicts the TMC-Regions in the morning and evening, as well as the k-means significant regions. Looking at the maps in Figure 12.9, we see from the TMC-Pattern extraction result that users spent time at the Chinese Military Museum in the morning (green box). However, and importantly, users do not visit the Chinese Military Museum in the evening. Such information on real dataset highlights TMC-Pattern's ability to detect interesting locations at different granularity.

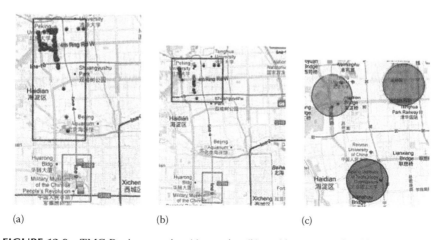

FIGURE 12.9 TMC-Region morning (a), evening (b), and k-means region (c).

12.4.3 FREQUENT PATTERNS

To evaluate the ability of TMC-Pattern to discover frequent patterns, we inputted multiuser TMC-Regions created by our extraction technique into our customized PrefixSpan algorithm (Algorithm 12.2). We should note that during the creation of the latter TMC-Region, only space and time were chosen as the relevant dimensions. We carried a second experiment using the Twitter–GeoLife Disease dataset to find ROI that are based on the *flu* disease context rather than location. This is accomplished by first passing the flu dataset into our extraction algorithm and choosing space (longitudes and latitudes) and disease as the relevant dimensions. This leads to the clustering of points based on the latter dimensions. The disease TMC-Regions are then passed onto Algorithm 12.2. Figure 12.10 illustrates the results of both experiments for different values of minimum support ranging from 0.2 to 0.6. The figure compares our TMC-Region ROI with the ROI generated by T-Pattern (Giannotti et al. 2007). With respect to location, our technique produces a better result, since it detects more frequent patterns than T-Pattern for the same minimum support. T-Pattern could not detect any ROI for the Twitter–GeoLife dataset because its trajectory model was tailored to handle only location and time. Our technique found 24 disease-related ROI from the Twitter search API when using the China-based GeoLife dataset. Figure 12.10 also shows that as the support increases, the number of frequent locations and disease patterns decreases.

12.4.4 LOCATION PREDICTION

We utilized only single-user trajectory histories for location prediction. Section 12.3.2.4 described how scores emanating from TMC-Pattern are combined with Markov model to predict the next location of a user. We extracted and utilized single-user TMC-Pattern from the GeoLife dataset and the Twitter–GeoLife

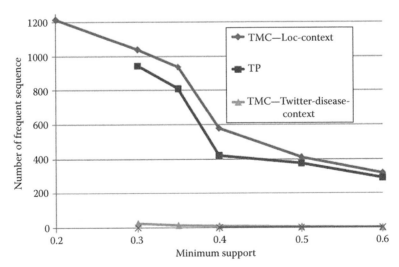

FIGURE 12.10   Frequent pattern.

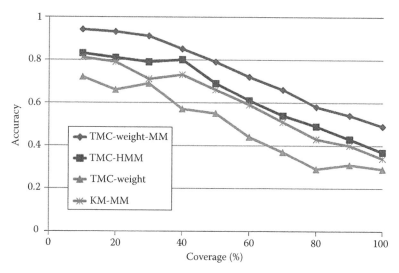

FIGURE 12.11 Prediction accuracy.

Disease dataset for prediction. In order to evaluate the effects of utilizing TMC-Pattern scores or weights during the experiments, apart from coupling scores from TMC-Pattern to a second-order Markov chain during prediction, we created two other predictive models. The first utilizes only a second-order Markov chain, while the second employs only scores from TMC-Pattern to predict the next location. In addition, we compared our next location predictor with the prediction technique proposed by Ashbrook and Starner (2003). We evaluated our prediction results with the use of the accuracy metric as shown in Figures 12.11 and 12.12. The accuracy and coverage are given by the following equations:

$$\text{Accuracy} = \frac{\text{Number of correct predictions}}{\text{Total number of TMC prediction}} \tag{12.2}$$

$$\text{Coverage} = \frac{\text{Number of TMC-patterns}}{\text{Total number of records in dataset}} \tag{12.3}$$

The result demonstrates that the combination of TMC-Pattern scores and second-order Markov model yields the best results. The latter combination has an accuracy of 94% for smaller datasets that is an 11% increase in prediction accuracy when using only Markov chain and 14% increased prediction accuracy when compared to UGLSL (i.e., KM-MM in Figure 12.11). This increase of prediction accuracy is credited to the use of TMC-Pattern scores.

The prediction result of the Twitter flu dataset, whose goal is to predict the next location of the spread of the flu disease, is depicted in Figure 12.12. Our technique achieved an 81% accuracy for a coverage of 20% for TMC-Regions with time period $T = \{M\}$, while for TMC-Regions in the evening ($T = \{E\}$), an accuracy of 73% is garnered for the same coverage. We observed that the accuracy decreases as the

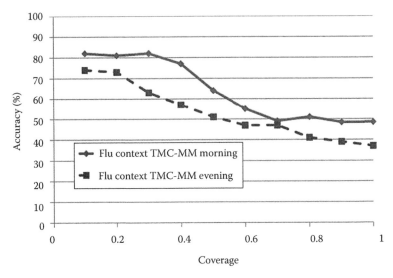

FIGURE 12.12 Disease spread prediction.

coverage increases. We further evaluated the impact of some TMC-Pattern vectors such as time period and residence mode during location prediction.

We performed this analysis by ignoring each of the aforementioned vectors during the prediction of a user's next location. Figure 12.13 illustrates this result that shows the cumulative probability of successful prediction for different observations. The figure illustrates that the TMC-Pattern score from the residence mode and time period are more important than those from the other vectors.

Besides, since lots of raw GPS points that are normally discarded by other techniques are incorporated into our trajectory model, we investigated if this leads to the creation

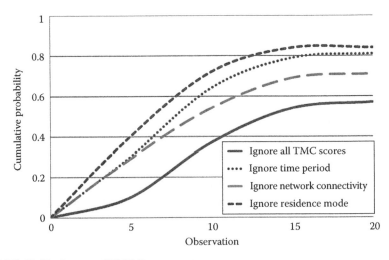

FIGURE 12.13 Impact of TMC-Pattern vectors.

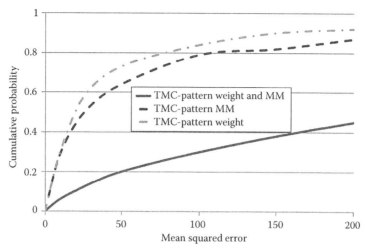

FIGURE 12.14   Prediction error.

of many outliers in the dataset and if the dataset is predictable. This can be tested by determining the ratio of the cumulative probability of the mean square error $\bar{E}$, since $\bar{E}$ is known (Han and Kamber 2005) for its ability to overestimate the presence of outliers:

$$\bar{E} = \frac{\sum_{j=1}^{t} \left(x_j - x_j' \right)^2}{t} \tag{12.4}$$

where x and x' are the true and predicted positions.

The outcome of this experiment is shown in Figure 12.14. Higher values of the cumulative probability mean greater chances of making error that might reflect the presence of outliers. The figure shows that the probability of making a false prediction for a large radius of 100 km is below 50% when TMC-Pattern is coupled with Markov chain. This indicates that there are less outliers in our model.

12.5   SUMMARY

We proposed a novel trajectory model called TMC-Pattern that can be exploited to create value from big geospatial data. TMC-Pattern rigorously captures and embeds infrastructural, human, social, and behavioral patterns. Most state-of-the-art trajectory models are tailored to model only spatiotemporal dimensions while mobility data emanating from state-of-the-art ubiquitous devices entail dimensions beyond these common dimensions. Our proposed TMC-Pattern handles a wide variety of dimensions and utilizes SUBCLU to find contextual ROI. In addition, we presented a new technique that utilizes our customized variant of the PrefixScan algorithm to uncover and project frequent patterns hidden in big geospatial data. We also showed through empirical experiments how TMC-Pattern can be employed to discover frequent locations or contexts associated with mobility such as disease spread or message context in social networks. Furthermore, we proposed a novel location predictor that employs TMC-Pattern, SUBCLU, and Markov chain. Our experiments demonstrate significant increases in prediction accuracy when TMC-Pattern is utilized.

REFERENCES

Asahara, A., K. Maruyama, A. Sato, and K. Seto. Pedestrian-movement prediction based on mixed Markov-chain model. *Proceedings of the 19th ACM SIGSPATIAL International Conference on Advances in Geographic Information Systems*, Chicago, IL. 2011. pp. 25–33.

Ashbrook, D. and T. Starner. Using GPS to learn significant locations and predict movement across multiple users. *Personal Ubiquitous Computing*. 2003. 7(5):275–286.

Eagle, N. and A. Pentland. Reality mining: Sensing complex social systems. *Personal Ubiquitous Computing*. 2006. 10(4):255–268.

Giannotti, F., M. Nanni, F. Pinelli, and D. Pedreschi. Trajectory pattern mining. *Proceedings of the 13th ACM SIGKDD International Conference on Knowledge Discovery and Data Mining*, San Jose, CA. 2007. pp. 330–339.

Han, J. and M. Kamber. *Data Mining: Concepts and Techniques*. Morgan Kaufmann, San Francisco, CA, 2005.

Han, S.-J. and S.-B. Cho. Predicting user's movement with a combination of self-organizing map and Markov model. *Proceedings of the 16th International Conference on Artificial Neural Networks—Volume Part II*, Athens, Greece. 2006. pp. 884–893.

Jungkeun, Y., B.D. Noble, L. Mingyan, and K. Minkyong. Building realistic mobility models from coarse-grained traces. *Proceedings of the Fourth International Conference on Mobile Systems, Applications and Services*, Uppsala, Sweden. 2006. pp. 177–190.

Kailing, K., H.-P. Kriegel, and P. Kröger. Density-connected subspace clustering for high-dimensional data. *Proceedings of the Fourth SIAM International Conference*, Lake Buena Vista, FL. 2004. pp. 246–257.

Kalasapur, S., H. Song, and D. Cheng. Extracting co-locator context. *Sixth Annual International Conference on Mobile and Ubiquitous Systems: Computing, Networking and Services*, Toronto, Canada. 2009.

Kitsak, M., L. Gallos, S. Havlin, F. Liljeros, and H. Makse. Identification of influential spreaders in complex networks. *Nature Physics*. 2010. 6:888–893.

Laasonen, K., M. Raento, and H. Toivonen. Adaptive on-device location recognition. *Pervasive Computing*. 2004. 3001:287–304.

Liao, L., D. Fox, and H. Kautz. Extracting places and activities from GPS traces. *International Journal of Robotics Research*. 2007. 26(1):119–134.

Minkyong, K., D. Kotz, and K. Songkuk. Extracting a mobility model from real user traces. *Proceedings of the 25th Annual Joint Conference of the IEEE Computer and Communications Societies (INFOCOM)*, Barcelona, Spain. IEEE Computer Society Press, Uppsala, Sweden, 2006.

Monreale, A., F. Pinelli, R. Trasarti, and F. Giannotti. Where next: A location predictor on trajectory pattern mining. *Proceedings of the 15th ACM SIGKDD International Conference on Knowledge Discovery and Data Mining*, Paris, France. 2009. pp. 637–646.

Morzy, M. Prediction of moving object location based on frequent trajectories. *Proceedings of the 21st International Conference on Computer and Information Sciences*, Milwaukee, WI. 2006. pp. 583–592.

Nicholson, A.J. and B.D. Noble. BreadCrumbs: Forecasting mobile connectivity. *Proceedings of the 14th ACM International Conference on Mobile Computing and Networking (MobiCom)*, San Francisco, CA. 2008. pp. 46–57.

Obasi, G.O.P., 1994; WMO's role in the international decade for natural disaster reduction. *Bull. Amer. Meteor. Soc.*, 75, 1655–1661. doi: http://dx.doi.org/10.1175/1520-0477 (1994)075<1655:WRITID>2.0.CO;2

Pei, J. et al. PrefixSpan: Mining sequential patterns by prefix-projected growth. *Proceedings of the 17th International Conference on Data Engineering (ICDE)*, Heidelberg, Germany. IEEE Computer Society, 2001.

Quannan, L., Y. Zheng, X. Xing, C. Yukun, L. Wenyu, and W.-Y. Ma. Mining user similarity based on location history. *Proceedings of the 17th ACM SIGSPATIAL International Conference on Advances in Geographic Information Systems*. 2008. 34:1–34, 10.

Yavas, G., D. Katsaros, and Y. Manolopoulos. A data mining approach for location prediction in mobile environments. *Data & Knowledge Engineering*. Elsevier Science Publishers B.V, Amsterdam, the Netherlands, Vol. 54, 2005. 121–146.

Yu, Z., Z. Lizhu, X. Xing, and W.-Y. Ma. Mining correlation between locations using human location history. *Proceedings of the 17th ACM SIGSPATIAL International Conference on Advances in Geographic Information Systems*, Seattle, WA. 2009. pp. 472–475.

13 Geospatial Cyberinfrastructure for Addressing the Big Data Challenges on the Worldwide Sensor Web

Steve H.L. Liang and Chih-Yuan Huang

CONTENTS

13.1 INTRODUCTION

In the next century, planet earth will don an electronic skin. It will use the Internet as a scaffold to support and transmit its sensations. This skin is already being stitched together. It consists of millions of embedded electronic measuring devices; thermostats, pressure gauges, pollution detectors, cameras, microphones, glucose sensors, EKGs, electroenceph-alographs. These will probe and monitor cities and endangered species, the atmosphere, our ships, highways and fleets of trucks, our conversations, our bodies-even our dreams.

Neil Gross
"The Earth Will Don an Electronic Skin," Businessweek, 1999

In recent years, large-scale sensor arrays and the vast datasets they produce worldwide are being utilized, shared, and published by a rising number of researchers on an ever-increasing frequency. Examples include the global-scale Argos network of buoys,* the weather networks of the World Meteorological Organization, and the global GPS zenith total delay (ZTD) observation network. Significant amount of efforts (e.g., GEOSS[†] and NOAA IOOS[‡]) have been put forth to web-enable these large-scale sensor networks so that the sensors and their data can be accessible through interoperable sensor web standards. Moreover, with the advent of the low-cost sensor networks and data loggers, it is technologically and economically feasible for individual scientists to deploy and operate small- to medium-scale sensor arrays at strategic locations for their own research purposes. There is a spectrum of sensor networks ranging from local-scale short-term sensor arrays to global-scale permanent observatories. The vision of the *worldwide sensor web* is becoming a reality.

The original worldwide sensor web concept was proposed by the NASA/Jet Propulsion Laboratory (JPL) in 1997 (Delin 2005; Liang et al. 2005) for acquiring the environmental information by integrating massive spatially distributed consumer-market sensors. The sensor web/network is increasingly attracting interest of researchers for a wide range of applications, including large-scale monitoring of the environment (Hart and Martinez 2006), civil structures (Xu et al. 2002), roadways (Hsieh 2004), and animal habitats (Mainwaring et al. 2002). Ranging from video camera networks that monitor real-time traffic to matchbox-sized wireless sensor networks embedded in the environment to monitor habitats, the worldwide sensor web generates tremendous volumes of priceless streaming data that enable scientists to observe previously unobservable phenomena.

Similar to the World Wide Web (WWW), which acts essentially as a *worldwide computer*, the sensor web can be considered as a *worldwide sensor* or a *cyberinfrastructure* that instruments and monitors the physical world at spatial and temporal scales that were previously impossible. However, realizing the sensor web vision is challenging. In order to construct a sensor web system, *the big data challenges on the worldwide sensor web* need to be addressed.

13.2 BIG DATA CHALLENGES ON THE WORLDWIDE SENSOR WEB

In the context of information technology (IT), the term *big data* is used to describe datasets that are too large or too complex to manage and process with traditional database management systems and data processing applications. The most widely recognized model for big data is *the 3Vs model*, which was first used by Doug Laney in 2001 (Laney 2001). While the 3Vs represent the *volume, velocity,* and *variety,* the 3Vs model defines the big data as the data with large volume, large velocity, and large variety. And each V would pose different data management challenges.

---

* http://www.argos-system.org
† http://www.earthobservations.org/geoss.shtml
‡ http://ioos.gov/

With the rapid development of the sensor web, we have observed that the sensor web also encounters the big data challenges. We explain the basic concept of the 3Vs model and how the sensor web fits into this model as follows.

Volume: One characteristic of big data is the large data volume, which can mean the large size or the large number of data records. While the data volume in social media is known as enormous (e.g., Facebook.com has more than 500 TB of new data every day), Stephen Brobst, the CTO of Teradata, in 2010 predicted (Kwang 2010) that "I don't think social media will be the biggest store of unstructured data for long ... Within the next three to five years, I expect to see sensor data hit the crossover point ... From there, the former will dominate by factors; not just by 10 to 20 percent, but by 10 to 20 times that of social media." It is foreseeable that with the increasing number of sensors being deployed worldwide, the sensor web will be generating more and more sensor data every day. Therefore, how to store, transmit, and process sensor data in large volume is one of the major challenges on the sensor web.

Velocity: The velocity characteristic refers to the rate at which data is produced. Unlike the human participants in the social media, the sensors in the sensor web can produce data in very high frequencies as long as they have enough power supply. For example, the Boeing jet engines produce 10 TB of sensor data every 30 min during the flight (Rogers 2011). Other examples are the closed-circuit television cameras (CCTV) and Internet protocol (IP) cameras for traffic monitoring or surveillance purposes that produce pictures from few to 30 frames per second. As a result, how to efficiently process the high-velocity sensor data streams is challenging, especially if we consider the geospatial nature of sensor data.

Variety: Many examples have been used to explain the variety characteristic, including nonaligned data structures, inconsistent data semantics, and incompatible data formats (Laney 2001). In general, the variety characteristic refers to the fact that there are some differences between the data records. In the sensor web, although the sensor data are relatively structured in comparison to the data in the social media, the sensor data have large variety in terms of hardware (i.e., different types of sensors), data types (e.g., video, image, text, and number), observed phenomena (e.g., RGB, air temperature, wind speed), communication protocols (e.g., proprietary protocols), data encodings (e.g., XML, JSON, JPEG), semantics (e.g., the same term is interpreted differently), and syntaxes (e.g., the same concept is described differently). Therefore, how to effectively integrate the heterogeneous sensor data and provide a coherent view for users is one of the major challenges on the sensor web.

As now we understand the high-level concept of the big data challenges on the sensor web, we need to understand the current sensor web development to further analyze the data management issues caused by the big sensor data.

13.3 WORLDWIDE SENSOR WEB ARCHITECTURE

To the best of our knowledge on the current sensor web development, the architecture of the worldwide sensor web is very similar to that of the WWW. For example, the WWW connects every web services around the world together through open standard protocols (e.g., HTTP), which has been demonstrated very scalable in terms of interchanging messages worldwide. The current sensor web development is moving toward the similar direction. We can see this trend from many sensor web projects, such as SensorWare Systems,* Microsoft SenseWeb project,† and COSM. com.‡ These projects deploy sensors, collect sensor data, and host and share the data on the WWW through proprietary protocols. In addition, there are several standard working groups developing open standards for the sensor web services, such as the Open Geospatial Consortium (OGC) Sensor Web Enablement (SWE) standards (Botts et al. 2007).

Similar to the WWW, the sensor web mainly has three layers, namely, the data layer, the web service layer, and the application layer. The sensor web layer stack is shown in Figure 13.1. The data layer can be further divided into the physical layer and the sensor layer. While the data layer performs observations§ and transmits sensor data to the web service layer, the web service layer provides the access for the application layer to retrieve the cached sensor data.

Since the architectures of the sensor web and the WWW are very similar, the components that are essential for the WWW could be considered in the development of the sensor web. For example, in the following, we identify three high-level components that are essential for the current WWW, namely, the open standard protocols, the resource discovery services, and the client-side platforms.

Open Standard Protocols: The open standard protocols play the most important role in the success of the WWW. The open standard protocols handle the communications in the Internet layers, such as the seven layers in the OSI model (ISO/IEC 7498-1) and the four layers in the TCP/IP model (IETF RFC 1122). For example, the IEEE 802 standards¶ define protocols for the local area networks (LAN), including the Ethernet and the wireless LAN. The IP defines the format of Internet packets and provides an addressing system for routing packets from a source host to a destination host. The transmission control protocol (TCP) and user datagram protocol (UDP) are the two commonly used standards in the transport layer. Furthermore, the hypertext transfer protocol (HTTP) is a protocol in the application layer that controls the high-level communications between applications. For example, a user may use a web browser, a client-side application, to send an HTTP request to an application running on a server hosting a website. Then the server could return resources, such as hypertext markup language (HTML) files, in an HTTP response to the client.

These open standards (and many others) developed by the Internet Engineering Task Force (IETF), the WWW Consortium (W3C), and the ISO/International Electrotechnical Commission (IEC) are to make sure that the Internet components are interoperable, which significantly contribute to the success of the WWW. On the other hand, to prevent *reinventing the wheels*, the sensor web needs to be

---

* http://www.sensorwaresystems.com/
† http://research.microsoft.com/en-us/projects/senseweb/
‡ https://cosm.com/
§ Here we follow OGC SWE's definition of observation, which is "an act of observing a property or phenomenon, with the goal of producing an estimate of the value of the property."
¶ http://standards.ieee.org/about/get/802/802.html

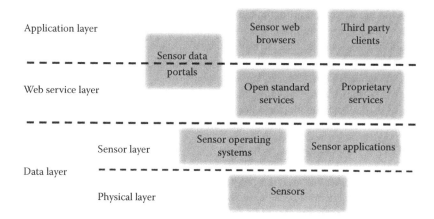

FIGURE 13.1   The sensor web layer stack.

built on top of many existing WWW standards. However, as the content on the sensor web is fundamentally different from that on the WWW, additional open standard protocols need to be defined. For example, the OGC community has developed the SWE standards, recognized as the most comprehensive sensor web standards framework. It can be considered as the key enabler for realizing the sensor web vision.

Resource Discovery Services: As the open standard protocols handle the communications between web services hosted worldwide, the resource discovery is a critical issue considering the highly distributed nature of the WWW. Today's WWW uses search engines (e.g., Google and Bing) and social bookmarking systems (e.g., Reddit.com and StumbleUpon.com) to address the resource discovery issue and to help users find the resources they are interested in. Similarly, the sensor web resource discovery issue must be addressed as well.

Client-Side Platforms: Client-side platforms allow users to send requests and visualize responses and have become an essential part of the WWW. The most popular type of client-side platforms is the web browsers (e.g., Internet Explorer, Chrome, and Firefox). As long as the web browsers and the web services follow the same protocols, users can use web browsers to communicate with web services (through HTTP) and visualize the response content (e.g., HTML documents, images, videos).

For the sensor web, we recognize that the most common client-side platforms are the *sensor data portals*, which serve as intermediaries between users and the sensor data they host. Since sensor data portals have full knowledge about the data they host (e.g., sensor locations and sampling times), they can pre-generate indices with the spatiotemporal distribution of sensor data to optimize data transmission. For example, Ahmad and Nath (2008) proposed the COLR-Tree to aggregate and sample sensor data to reduce data size before transmission. Some sensor data portals, for example, Groundwater Information Network (GIN),* present a map of sensor

---

* http://analysis.gw-info.net/gin/public.aspx

locations at small scale and actual sensor observations at large scale to limit the number of sensor observations being transmitted in each request. However, a critical drawback of these sensor data portals is that they can only present the sensor data about which they have some prior knowledge.

Instead, we envision that a client-side platform for the sensor web should be capable of communicating with any sensor web services that follow the same protocol. As this type of client-side platform does not require any prior knowledge of the data to retrieve and visualize sensor data, we call this pure client-side application *sensor web browser*.

Additional Components: Besides the previous three essential components, some other ideas on the WWW may be helpful for the sensor web as well. For example, the online social network (OSN) may be helpful to recommend sensor data to users according to their interests. In addition, as the Web 2.0 is one of the fundamental concepts of volunteered geographic information (VGI) (Goodchild 2007) and has demonstrated its usefulness (e.g., the OpenStreetMap), the sensor web can certainly benefit from this concept. Moreover, one of the important and ongoing directions of the WWW is the semantic web,* which aims to convert the unstructured and semi-structured content on the current WWW into a *web of data*. Although the semantic web still seems premature, we believe the sensor web can also apply the semantic web concept to integrate sensor data from different services.

As the sensor web development moves toward the architecture of the WWW, the distributed nature of web services would be an effective solution to address the issue of storing large volumes of sensor data. However, besides the storage issue, there are other big data challenges to be solved, such as transmitting large volumes of sensor data across the network, efficiently retrieving and updating high-velocity sensor data streams, and effectively integrating heterogeneous sensor data. In order to address these big sensor data challenges and realize the sensor web vision, we propose the Geospatial Cyberinfrastructure for Environmental Sensing (GeoCENS) architecture as a possible direction.

13.4   GEOCENS ARCHITECTURE

The GeoCENS architecture is built as an online platform for the sensor web. With GeoCENS, users can maneuver a sensor web browser, within a 3D virtual globe or on a 2D base map, to discover, visualize, access, and share heterogeneous and ubiquitous sensing resources and other relevant information. Our aim is to address the aforementioned technical challenges, propose innovative approaches, and provide the missing software components for realizing the worldwide sensor web vision.

Figure 13.2 shows the GeoCENS architecture. First, similar to that in the WWW, everyone can build and deploy sensor web services to host sensor data. As those sensor web services may distribute worldwide and are not registered on any catalogue service, a *sensor web service search engine* is proposed for GeoCENS to discover

---

* http://www.w3.org/standards/semanticweb/

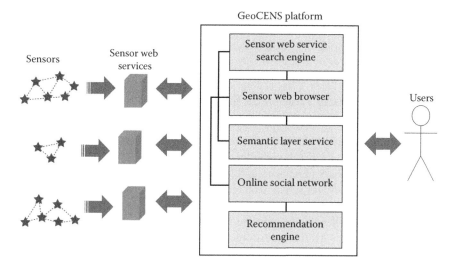

FIGURE 13.2   The GeoCENS architecture.

and index sensor web services and consequently to allow users to search sensor web services with query criteria. GeoCENS also features a *sensor web browser* for users to retrieve and visualize sensor data from sensor web services. In addition, the *semantic layer service* utilizes the metadata from sensor web services to integrate heterogeneous sensor data layers in order to provide users a coherent view of the sensor web data. Furthermore, GeoCENS has an *OSN* component that allows users to establish friendships and share sensor data. With the friendship links on the OSN, GeoCENS provides a *recommendation engine* that recommends sensors and datasets according to a user's interests.

13.4.1   OGC-Based Sensor Web Servers

GeoCENS uses the OGC SWE open standards as the fundamental interoperability architecture. The SWE is an OGC initiative that builds an open and interoperable geospatial web service framework to enable the exchange and processing of sensor observations and relevant sensor web data (Botts et al. 2007).

The availability of the sensor web service implementations is very important for sensor data owners to easily install and share their sensor data in an interoperable manner. For example, the 52° North Sensor Observation Service (SOS),* the MapServer SOS,† and the Deegree SOS‡ implementations are available for public to download and install. GeoCENS also implemented the OGC SOS specification (Na and Priest 2007), SensorML specification (Botts and Robin 2007), and Observation and Measurement (O&M) specification (Cox 2011). GeoCENS SOS implementation is unique in that the NoSQL database approach (e.g., the Apache

---

* http://52north.org/communities/sensorweb/sos/

† http://mapserver.org/ogc/sos_server.html

‡ http://wiki.deegree.org/deegreeWiki/deegree3/SensorObservationService

CouchDB and MongoDB) can be applied to implement the GeoCENS server, which may serve as a more scalable backend. The GeoCENS SOS implementation (CouchDB version) is also released online* to help sensor data owners *open* their data.

13.4.2 DECENTRALIZED HYBRID P2P SENSOR WEB SERVICE DISCOVERY

For any large-scale distributed system (e.g., the WWW), both communication and data management distill down to the problem of resource discovery. Similarly, GeoCENS needs a sensor web resource discovery service. In order to handle sensor web's large numbers of sensors and large numbers of users, GeoCENS uses a hybrid P2P architecture for sensor web resource discovery. Every GeoCENS sensor web server also serves as part of the sensor web service discovery infrastructure (i.e., a peer node). These nodes operate on a cooperative model, where each peer leverages each other's available resources (i.e., CPU, storage, and bandwidth) for mutual benefit.

From literature and existing systems, there are two types of P2P architectures, namely, unstructured P2P networks, for example, CAN (Ratnasamy et al. 2001), Pastry (Rowstron and Druschel 2001), and Chord (Stoica et al. 2001), and structured P2P networks, for example, Gnutella (Ripeanu et al. 2002). Unstructured P2P networks are networks where participating nodes perform actions for each other, where no rules exist to define or constrain connectivity between nodes. They are simple but not scalable because their flood-based query processing generates enormous amounts of network traffic. Structured P2P networks use hash functions to build distributed indexes for their stored data items. The hash tables, like distributed indexes, successfully reduce the nodes to be scanned per query. However, structured P2P networks are vulnerable to node dynamics.

A hybrid approach that uses both structured and unstructured P2P networks is proposed in GeoCENS (Chen and Liang 2011). The rationale for such a hybrid design is described as follows. We envision that the future sensor web will have two types of sensor web servers: (1) powerful sensor web servers maintained by large institutions (e.g., NASA or NOAA) (they would not join and leave the network randomly—in most cases these servers are made accessible 24/7, which means they are static nodes in the network); and (2) less powerful sensor web servers maintained by small institutions or even individuals (e.g., universities or citizen scientists) (these servers might join and leave the network more frequently and are dynamic and transient nodes in the network). Considering the previously described settings, it is a rational design decision to group static P2P nodes into structured super-nodes (to exploit the stability of static nodes) and group dynamic P2P nodes into leaf nodes (to save the overhead for maintaining the structure). Since structured P2P networks can only process exact key–value pair queries, we enable geospatial search functions by labeling data with space-filling curves (SFC). Our architecture is also unique in that it is a locality-aware system, that is, the system is able to exploit the locality information between peer nodes in order to deliver the query results efficiently.

---

* http://wiki.geocens.ca/sos

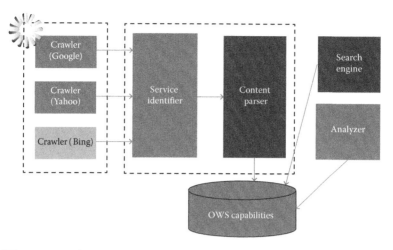

FIGURE 13.3   The GeoCENS search engine architecture.

In addition, for the non-GeoCENS OGC web services (OWS), which are not on the P2P overlay network, the GeoCENS search engine implements crawlers to periodically look for and index the online OWS services. Therefore, users can still be able to find these services through the GeoCENS search engine. The architecture of the GeoCENS search engine is shown in Figure 13.3. Currently, the GeoCENS search engine has discovered 2,884 web mapping services (WMS), which have 88,281 WMS layers, and 59 SOS services, which have 6,295 SOS observation offerings.

13.4.3   3D Virtual Globe-Based and 2D Map-Based Sensor Web Browsers

The GeoCENS sensor web browser is an intuitive 3D client front end for all OGC SOS services and OGC WMS (de la Beaujardiere 2006). It allows users to maneuver a 3D sensor web browser, within a single virtual globe, in order to browse, discover, visualize, access, share, and tag heterogeneous sensing resources and other relevant information. Starting from a *zoomed out* view of the globe, users are able to select a study site and *fly* into it. While flying to their study sites, multiresolution map data are loaded to the client from the WMS. The GeoCENS browser combines multisensor data streams and geographical datasets and renders them in a coherent and unified virtual globe environment.

We have developed the GeoCENS sensor web browser on top of the open source World Wind virtual globe system.* To the best of our knowledge, it is the world's first OGC-based sensor web 3D browser. The GeoCENS browser has the following two unique components/contributions: (1) in order to interoperate with existing sensor web servers, we have developed an OGC SWE communication module that is able to communicate with OGC SWE-compatible servers, and (2) in order to prevent

---

* http://worldwind.arc.nasa.gov/

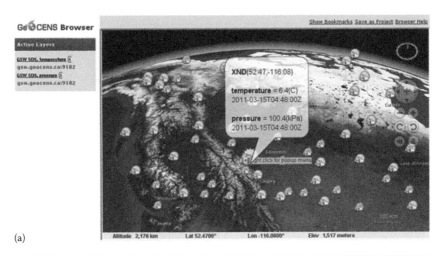

(a)

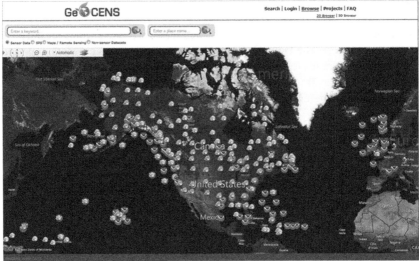

(b)

FIGURE 13.4   Screenshots of (a) 3D virtual globe-based and (b) 2D map-based sensor web browsers.

transferring large volumes of sensor data across the network repeatedly, we have developed a new spatiotemporal data loading module called LOST-Tree (Huang and Liang 2012b) with client-side cache. A screenshot of the 3D virtual globe-based sensor web browser is shown in Figure 13.4a.

In addition to the 3D virtual globe-based sensor web browser, GeoCENS also features a lightweight 2D map-based sensor web browser, which retrieves sensor data cache from a mediator called *translation engine* (Knoechel et al. 2011). As the translation engine handles the heavy communication load with the sensor web services, the 2D map-based sensor web browser can retrieve the cached sensor data from the translation engine in a lightweight and efficient manner, which is mobile friendly. A screenshot of the 2D map-based sensor web browser is shown in Figure 13.4b.

Moreover, in order to address the high-velocity big data issue, the translation engine utilizes an *adaptive feeder* that detects the data updating frequency on the sensor web services and fetches the latest sensor data from the services by adaptively scheduling the requests (Huang and Liang 2012a). In this case, the cached sensor data in the translation engine are always updated.

13.4.4 ONLINE SOCIAL NETWORK

GeoCENS is an OSN-based sensor web platform for researchers. Take Facebook. com, today's most popular OSN, as an example. On Facebook, users can share photos/ activities with their friends and networks. On GeoCENS, researchers can interact and share sensors, scientific datasets (including data from sensors), experiences, and activities with their friends (e.g., colleagues from other institutes) and networks. GeoCENS users can create a profile where they declare their research interests and preferences and establish friendships with other users. Figure 13.5 shows an example of the GeoCENS user profile. A *friendship* is formed on GeoCENS when one GeoCENS user extends a friendship invitation to another user. Upon confirmation

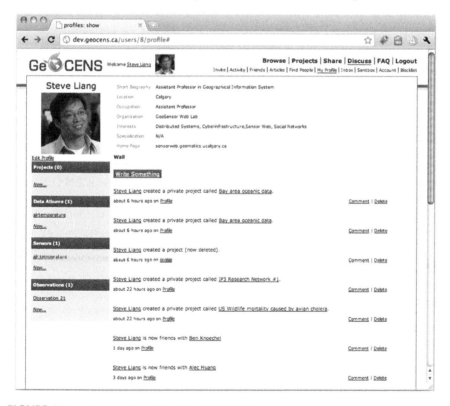

FIGURE 13.5   A screenshot of the GeoCENS OSN.

by the latter, the friendship relationship is formed. Other features include the ability to upload sensor datasets, the ability to join projects/groups of shared area of research interest, the ability to adjust different privacy levels, and the ability to review/annotate/rate sensors as well as datasets.

By creating a specialized OSN for sensor web users, our goal is to leverage the underlying social graphs, the structure of user interactions, and the users' profiles/ preferences to create innovative uses and applications of the sensor web. One innovative OSN-based sensor web application is a sensor web recommendation engine, which we will describe in the next section.

13.4.5 Sensor Web Recommendation Engine

With the GeoCENS social network infrastructure, we are able to develop a sensor web recommendation engine (i.e., a collaborative tagging system) that recommends sensors and datasets according to a user's geographical area of interest. In fact, existing folksonomy-related research is mostly focused on non-geospatial applications (Hotho et al. 2006). One key contribution of the GeoCENS recommendation engine is that we extend the folksonomy research into geospatial applications by leveraging the geospatial information associated with three key components of collaborative tagging systems, namely, tags, resources, and users.

GeoCENS's recommendation engine provides the following three functions: (1) Collaborative tagging, which enables users to assign tags to the resources (e.g., assign multiple tags to a sensor). In order to make the task easier for the user and to avoid ambiguity, our recommendation engine is able to suggest tags to the user. We have developed a new algorithm to suggest tags taking geospatial characteristics of the sensor web resources into consideration (Rezel and Liang 2011); (2) Collaborative browsing, which enables users to navigate through the tags collected in the system. It aids in the process of sensor and sensor data discovery. We have developed a new algorithm for building tag maps, a tag cloud for a sensor web browser where the geospatial attributes of the tag assignments are taken into consideration. Figure 13.6 shows a screen capture of an example GeoCENS tag map; and (3) Collaborative searching, which enables users to retrieve resources based on tag queries, either by clicking through a tag cloud or by typing the tag out. The key is to retrieve the most relevant results for these queries, and we have proposed an algorithm to enhance the query processing by taking geospatial aspects of the queries and data into consideration (Rezel and Liang 2011).

13.4.6 Semantic Layer Service

As the sensor web is highly heterogeneous in terms of hardware, data types, observed phenomena, communication protocols, data encodings, semantics, and syntactic, among other things, some of these heterogeneities can be addressed by applying the open standard protocols, such as the communication protocol and the data encoding heterogeneities. However, even after applying the same open standards, we found that there are interoperability problems from a lack of standardized naming, such as the semantic heterogeneity and the syntactic

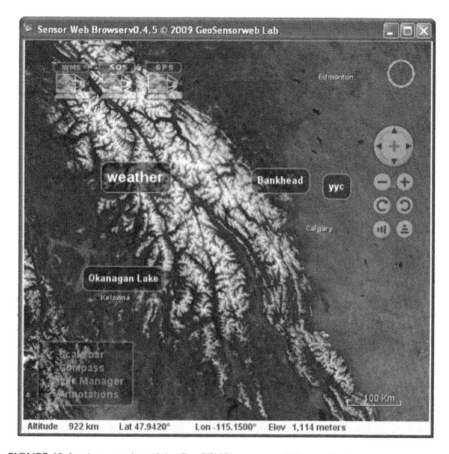

FIGURE 13.6   A screenshot of the GeoCENS recommendation engine tag map.

heterogeneity. For example, consider the two strings *precipitation* and *rainfall*. Since rainfall is a type of precipitation, a user interested in precipitation data would be interested in rainfall. Although these concepts are intuitively related to any human, to any computer these are simply different sequences of characters. On the other hand, different data providers will label their data differently. For example, Table 13.1 shows the various uniform resource identifiers (URIs) used in SOS services to represent the same concept of wind speed. This syntactic heterogeneity causes difficulties to design a system to integrate all sensor data layers that measure wind speed.

In order to address these heterogeneity issues in GeoCENS, a semantic layer service, which uses a bottom-up approach, including text processing, approximate string matching, and semantic string matching of data layers, to integrate sensor data layers by their phenomena (Knoechel et al. 2013), is proposed. The semantic layer service uses WordNet as a lexical database to compute word pair similarities and derive a set-based dissimilarity function using those scores. With the semantic layer service, we can effectively integrate the heterogeneous sensor data and provide a coherent view for users.

TABLE 13.1

Various URIs of the Concept of Wind Speed

1	urn:x-ogc:def:property:OGC::WindSpeed
2	urn:ogc:def:property:universityofsaskatchewan:ip3:windspeed
3	urn:ogc:def:phenomenon:OGC:1.0.30:windspeed
4	urn:ogc:def:phenomenon:OGC:1.0.30:WindSpeeds
5	urn:ogc:def:phenomenon:OGC:windspeed
6	urn:ogc:def:property:geocens:geocensv01:windspeed
7	urn:ogc:def:property:noaa:ndbc:Wind Speed
8	urn:ogc:def:property:OGC::WindSpeed
9	urn:ogc:def:property:ucberkeley:odm:Wind Speed Avg MS
10	urn:ogc:def:property:ucberkeley:odm:Wind Speed Max MS
11	http://marinemetadata.org/cf#wind speed
12	http://mmisw.org/ont/cf/parameter/winds

13.5 APPLICATIONS POWERED BY GEOCENS

As the GeoCENS project is designed for environmental researchers, we have also demonstrated that the GeoCENS architecture can be applied to many other applications. For example, the Rocky View County project* is a long-term ground-water monitoring project, where farmers can upload the water level data of their water wells (i.e., Web 2.0 and VGI). Currently, the Rocky View County project has more than 40 active water well owners, and the earliest data can be traced back to 2008. In addition, when the farmers upload their well readings, the hydrologists at the University of Calgary will receive notifications and perform the quality assurance (QA) and quality control (QC) processes. A screenshot of the Rocky View County project is shown in Figure 13.7a.

One other example is the Rocky Mountain Eagle Watch project,[†] which allows bird observers to submit the time and location they observed a bird. The Rocky Mountain Eagle Watch is also a citizen sensing (i.e., VGI) project for counting eagles for more than 20 years in the Rockies. In the past, the bird observers used papers and Excel data sheets to record and share their data, which is time-consuming and not efficient. By adapting the GeoCENS cyberinfrastructure architecture, the Rocky Mountain Eagle Watch platform has significantly simplified the data entry, analyzing, and sharing processes. A screenshot of the Rocky View County project is shown in Figure 13.7b.

These *powered-by-GeoCENS* applications are also open to public and have demonstrated that the GeoCENS architecture can successfully address the big sensor data challenges and consequently realize the worldwide sensor web vision.

---

* http://rockyview.geocens.ca/
† http://eaglewatch.geocens.ca/

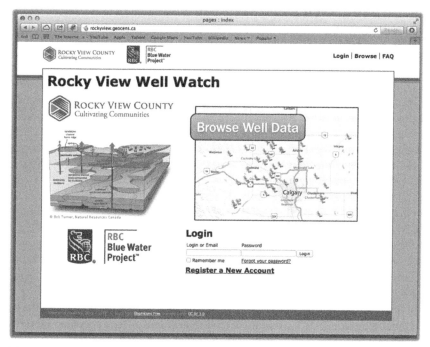

(a)

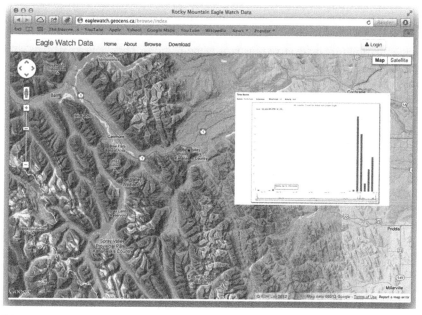

(b)

FIGURE 13.7 Screenshots of (a) Rocky View County and (b) Rocky Mountain Eagle Watch projects.

13.6 RELATED WORKS

Several recent works have attempted to propose architectures for sensor web systems. Intel Research's IrisNet (Gibbons et al. 2003) proposes a decentralized architecture based on a hierarchical topology and provides techniques to process queries over a distributed XML document containing sensor data. However, IrisNet only supports very preliminary geospatial queries. It uses hierarchical place names to build its hierarchical network topology. In order to perform geospatial query, users/applications need to know the exact place name a priori and explicitly specify the parts of a hierarchy that the query needs to traverse. Moreover, IrisNet does not have a sensor discovery module and can only handle homogeneous sensors and data types.

Microsoft Research's SensorMap (Ahmad and Nath 2008) uses a centralized web portal design and tackles the scalability and performance issues by building the COLR-Tree, a data structure that indexes, aggregates, and caches sensor streams, in order to prevent transferring large volume of sensor streams across the network. However, SensorMap's centralized design makes the portal a single point of failure. In contrast, GeoCENS uses a service-oriented architecture (SOA) and a decentralized hybrid P2P architecture for sensor web service discovery. There is no single point of failure in GeoCENS, and the P2P sensor discovery service balances the load by directing queries and traffic to the distributed sensor web services. Moreover, both IrisNet and SensorMap are based on proprietary interfaces and sensor data encodings, while GeoCENS follows OGC SWE specifications and is able to interoperate with other OGC-compliant sensor web servers.

GeoCENS is innovative and unique in that it is a social network-based sensor web platform. GeoCENS harvests the sensor web users' interaction structures and activities in order to build innovative sensor web applications. For example, with its social network infrastructure, GeoCENS is able to build a geospatial folksonomy for the sensor web that recommends relevant sensor web resources to a user according to the collective intelligence of the GeoCENS users.

13.7 SUMMARY

The worldwide sensor web has become a very useful technique on monitoring the physical world at spatial and temporal scales that were previously impossible. However, we believe that the full potential of the worldwide sensor web is not yet realized. In this chapter, we analyzed the challenges and the necessary components that are required to be developed to realize the vision of the sensor web. We first identified the big data challenges on the worldwide sensor web, which followed by the introduction of the sensor web architecture and its components. Then we introduced the GeoCENS architecture and the high-level concept on each component as one possible solution to address the challenges of realizing the sensor web vision. In addition, we also presented some powered-by-GeoCENS projects to demonstrate how the GeoCENS architecture can successfully address the big sensor data challenges and consequently realize the worldwide sensor web vision.

ACKNOWLEDGMENTS

The authors would like to thank CANARIE, Cybera, Alberta Innovates Technology Futures, and Microsoft Research for their supports on this project.

REFERENCES

Ahmad, Y. and S. Nath. 2008. COLR-Tree: Communication-efficient spatio-temporal indexing for a sensor data web portal. In *IEEE International Conference on Data Engineering*, Cancún, México.

Botts, M., G. Percivall, C. Reed, and J. Davidson. 2007. OGC sensor web enablement: Overview and high level architecture (OGC 07-165). Open Geospatial Consortium white paper, December 28, 2007.

Botts, M. and A. Robin. 2007. OpenGIS Sensor model language (SensorML) Implementation Specification (OGC 07-000). OGC Implementation Specification, July 17, 2007.

Chen, S. and S. Liang. 2011. A hybrid peer-to-peer architecture for global geospatial web service discovery. In *Spatial Knowledge and Information*, Fernie, British Columbia, Canada.

Cox, S. 2011. Observations and measurements—XML implementation (OGC 10-025r1). OGC Implementation, March 22, 2011.

de la Beaujardiere, J. 2006. OpenGIS Web map server implementation specification (OGC 06-042). OGC Implementation specification, March 15, 2006.

Delin, K. 2005. Sensor webs in the wild. In *Wireless Sensor Networks: A Systems Perspective*, Artech House, London, U.K.

Gibbons, P.B., B. Karp, Y. Ke, S. Nath, and S. Seshan. 2003. IrisNet: An architecture for a world-wide sensor web. *IEEE Pervasive Computing* 2(4):22–33.

Goodchild, M.F. 2007. Citizens as sensors: The world of volunteered geography. *GeoJournal* 69(4):211–221.

Gross, N. 1999. The earth will don an electronic skin. Online article, *Businessweek*. http://www.businessweek.com/1999/99_35/b3644024.htm (accessed March 27, 2013).

Hart, J.K. and K. Martinez. 2006. Environmental sensor networks: A revolution in the earth system science? *Earth Science Reviews* 78:177–191.

Hotho, A., R. Jaschke, C. Schmitz, and G. Stumme. 2006. Information retrieval in folksonomies: Search and ranking. *Lecture Notes in Computer Science* 4011:411.

Hsieh, T.T. 2004. Using sensor networks for highway and traffic applications. *IEEE Potentials* 23(2):13–16.

Huang, C.Y. and S. Liang. 2012a. A hybrid pull–push system for near real-time notification on sensor web. In *The XXII Congress of the International Society for Photogrammetry and Remote Sensing*, August 25–September 1, Melbourne, Victoria, Australia.

Huang, C.Y. and S. Liang. 2012b. LOST-Tree: A spatio-temporal structure for efficient sensor data loading in a sensor web browser. *International Journal of Geographical Information Science* 27:1190–1209.

Knoechel, B., C.Y. Huang, and H.L.S. Liang. 2011. Design and implementation of a system for the improved searching and accessing of real-world SOS services. *International Workshop on Sensor Web Enablement 2011*, October 6–7, Banff, Alberta, Canada.

Knoechel, B., C.Y. Huang, and S. Liang. 2013. A bottom-up approach for automatically grouping sensor data layers by their observed property. *ISPRS International Journal of Geo-Information* 2(1):1–26.

Kwang, K. 2010. Sensor data is data analytics' future goldmine. Online article, ZDNet. http://www.zdnet.com/sensor-data-is-data-analytics-future-goldmine-2062200657/ (accessed November 1, 2013).

Laney, D. 2001. 3D data management: Controlling data volume, velocity and variety. Gartner. http://blogs.gartner.com/doug-laney/files/2012/01/ad949-3D-Data-Management-Controlling-Data-Volume-Velocity-and-Variety.pdf (accessed March 27, 2013).

Liang, S.H.L., A. Croitoru, and C.V. Tao. 2005. A distributed geospatial infrastructure for sensor web. *Computers and Geosciences* 31(2):221–231.

Mainwaring, A., J. Polastre, R. Szewczyk, D. Culler, and J. Anderson. 2002. Wireless sensor networks for habitat monitoring. In *2002 ACM International Workshop on Wireless Sensor Networks and Applications*, Atlanta, GA.

Na, A. and M. Priest. 2007. OpenGIS Sensor observation service (OGC 06-009r6). OGC Implementation Standard, October 26, 2007.

Ratnasamy, S., P. Francis, M. Handley, R. Karp, and S. Schenker. 2001. A scalable content addressable network. In *Proceedings of the 2001 Conference on Applications, Technologies, Architectures, and Protocols for Computer Communications*, San Diego, CA, pp. 161–172.

Rezel, R. and S.H.L. Liang. 2011. A folksonomy-based recommendation system for the sensor web. *Proceedings of Web and Wireless Geographical Information Systems* 6574:64–77.

Ripeanu, M., A. Iamnitchi, and I. Foster. 2002. Mapping the Gnutella network. *IEEE Internet Computing* 6(1):50–57.

Rogers, S. 2011. Big data is scaling BI and analytics. Online article. Information Management. http://www.information-management.com/issues/21_5/big-data-is-scaling-bi-and-analytics-10021093-1.html (accessed March 27, 2013).

Rowstron, A. and P. Druschel. 2001. Pastry: Scalable, decentralized object location, and routing for large-scale peer-to-peer systems. *Lecture Notes in Computer Science* 2218:329–350.

Stoica, I., R. Morris, D. Karger, M.F. Kaashoek, and H. Balakrishnan. 2001. Chord: A scalable peer-to-peer lookup service for Internet applications. In *Proceedings of the 2001 Conference on Applications, Technologies, Architectures, and Protocols for Computer Communications*, San Diego, CA, pp. 149–160.

Xu, N. 2002. A survey of sensor network applications. *IEEE Communications Magazine* 40:102–144.

14 OGC Standards and Geospatial Big Data

Carl Reed

CONTENTS

14.1 WHAT IS THE OGC?

Founded in 1994, the Open Geospatial Consortium* (OGC) is a global industry consortium with a vision to "Achieve the full societal, economic and scientific benefits of integrating location resources into commercial and institutional processes worldwide." Inherent in this vision is the requirement for geospatial standards and strategies to be an integral part of business processes.

As of April 2013, the OGC had 475+ members—geospatial technology software vendors, systems integrators, government agencies, and universities—participating in a consensus process to develop, test, and document publicly available geospatial interface standards and encodings for use in information and communications industries. Open interfaces and protocols defined by OGC standards are designed to support interoperable solutions that *geo-enable* the web, wireless, and location-based services and mainstream IT and to empower technology developers to make complex spatial information and services accessible and useful to all kinds of applications.

---

* http://www.opengeospatial.org/.

As such, the mission of the OGC is to serve as a global forum for the development, promotion, and harmonization of open and freely available geospatial standards. Therefore, the OGC has a major commitment to collaborate with other standards development organizations that have requirements for using location-based content, such as the International Organization for Standardization (ISO), Organization for the Advancement of Structured Information Standards (OASIS), Internet Engineering Task Force (IETF), National Emergency Number Association (NENA), and Open Mobile Alliance (OMA).

14.2 INTRODUCTION

Massive amounts of geospatially enabled data are being collected every day. Further, the volume of geospatially enabled data collected continues to increase at a rapid rate. Every 24 h, unmanned aerial vehicles (UAVs) with sensor systems, satellite constellations, smart grid applications, GPS-enabled crowdsourced and social media applications, wide area motion imagery (WAMI), and many other collection systems are capturing, processing, and archiving petabytes of geospatially referenced content. To illustrate the problem of processing massive amounts of geo-enabled data, consider one example of the use of WAMI, as derived orthorectified image maps. One 150 megapixel color WAMI image frame at 8 bits per band uncompressed is about 450 MB. At a 10:1 compression, the storage requirement is 45 MB/frame. At 2 frames per second, this is 86,000 frames sampled during a 12 h period or about 3.7 TB of georeferenced imagery per day per sensor (OGC 2012).

Do the three Vs of high *volume*, high *velocity*, and/or high *variety* commonly used to characterize different aspects of big data apply to georeferenced data? There is no question, based on the one simple example provided earlier, that the geospatial community generates big data (volume). The velocity of collection can be incredible, gigabytes per second. Many real-time imaging sensors capture georegistered images at 1 or 2 frames per second. Another example of velocity occurs when a network of sensors feeds a continuous stream of real-time observations to an application. These observations could be wind speed, temperature, humidity, precipitation, and other phenomenon being captured by hundreds of sensors that are integrated into a network. Finally, the variety of geospatially enabled data is huge. From digitized maps to orthoimagery to survey data to civil engineering data to building information to georegistered photographs, the list is seemingly endless.

A key aspect of the OGC standards work related to big data is analytics, specifically what the OGC terms data fusion. Data fusion in the OGC is defined as "the act or process of combining or associating data or information regarding one or more entities considered in an explicit or implicit knowledge framework to improve one's capability (or provide a new capability) for detection, identification, or characterization of that entity." Location is the great data integrator. Fusion based on integrating disparate data sources using the geographic context provides a powerful basis for geospatially based big data analytics and predictive modeling.

Establishing a framework of open standards for geospatial data fusion environments will enable functional and programmatic developments for fusion beyond the current state of data fusion. Functional expansion will include fusion in manual and semiautomated multisource visualization and analysis as a preferred practice among analysts. Open standards will enable multiple fusion R&D projects with diverse technical approaches to collaborate. R&D prototypes can be enhanced for improved operational capability including access to data, integration of diverse inputs, and value-added applications. The OGC recognizes three types of fusion: feature, decision, and observation (sensor). Since vast amounts of geospatial big data are and will continue to be generated by location-enabled observation (sensor) systems, let's focus on that type of fusion.

14.3   WHAT IS OBSERVATION (SENSOR) FUSION?

Observation fusion considers sensor measurements of various observable properties to well-characterized observations including uncertainties. Fusion processes involve merging of multiple sensor measurements of the same phenomena (i.e., events of feature of interest) into a combined observation and analysis of the measurement signature. Sensor fusion concerns the acquisition and exploitation of multiple measurements for the purpose of

- Obtaining a higher-level or more accurate measurement
- Recognizing objects and events of interest
- Determining properties of particular objects or events
- Change detection
- Modeling, simulation, and alerting
- Counterterrorism
- Public safety
- Environmental monitoring

A recent example of sensor fusion is a European Seventh Framework Programme (FP 7) project titled Citizens OBservatory WEB (COBWEB). COBWEB is funded under the European Union's (EU) FP 7 in the Environment "Developing community-based environmental systems using innovative and novel earth observations applications" theme.

14.4   OGC STANDARDS AND BIG DATA

Numerous OGC standards have already been used for visualizing and processing location-enabled big data. These include the OGC Web Map Service (WMS) Interface, Web Coverage Service (WCS), Sensor Observation Service (SOS), and Web Processing Service (WPS). Further, a new candidate OGC standard, OpenMI, has a significant role in the future use of OGC standards for modeling big data. Figure 14.1 provides a simple architectural diagram of how these OGC standards are related and how they fit into a web services framework.

Each of these standards with an example is now described.

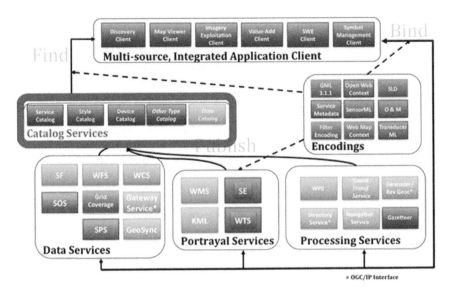

FIGURE 14.1 OGC standards and the web services paradigm.

14.4.1 OGC WMS Interface Standard

The OGC® WMS Interface Standard provides a simple HTTP interface for requesting georegistered map images from one or more distributed geospatial databases storing any type of geospatial data, such as raster images or features represented by vectors. A WMS request defines the geographic layer(s) and area of interest to be processed. The response to the request is one or more georegistered map images (returned as JPEG, PNG, etc.) that can be displayed in a browser application. The interface also supports the ability to specify whether the returned images should be transparent so that layers from multiple servers can be combined or not. Since the WMS is a visualization interface, its use is independent of the size of the big data repository. Using the same server performance would be the same for a request to a one terabyte database as it would to a petabyte database. As examples, this is why the US National Aeronautics and Space Administration (NASA) and the National Oceanographic and Atmospheric Administration (NOAA) use WMS for providing a simple visualization mechanism to many of their repositories. The NOAA Unified Access Framework for Gridded Data (UAF Grid) project implements WMS. There are a number of interoperability requirements driving the development and use of UAF Grid. These provide the ability to access and integrate location-enabled data from multiple archives and real-time sensor assets and using standards to access, deliver, and visualize these heterogeneous data assets to the end user. Another example is the use of WMS in NASA's Earth Observing System Data and Information System (EOSDIS)* Global Imagery Browse Services (GIBS).†

---

* https://earthd ata.nasa.gov/about-eosdis.
† https://earthdata.nasa.gov/about-eosdis/system-description/global-imagery-browse-services-gibs.

The GIBS provides a set of standard services to deliver global, full-resolution satellite imagery in a highly responsive manner. The goal of GIBS is to enable interactive exploration of the petabytes of NASA's Earth imagery for a broad range of users. Currently, 50+ full-resolution global imagery products from NASA's Earth Observing System satellites are available in near real time (usually within 3 h of observation).

14.4.2 OGC WCS INTERFACE STANDARD

The OGC® WCS defines a standard interface and operations that enables interoperable access to geospatial *coverages*. The term *grid coverages* typically refers to content such as satellite images, digital aerial photos, digital elevation data, and other phenomena represented by values at each measurement point. Services implementing these interfaces provide access to original or derived sets of geospatial coverage information, in forms that are useful for client-side rendering, input into scientific models, and other client applications. The response to a WCS request, delivered via HTTP, includes coverage metadata and an output coverage whose pixels are encoded in a specified binary image format, such as GeoTIFF or NetCDF.

An example of the use of WCS for big data access and analytics is the EU project *EarthServer.** The EarthServer project, initiated by Jacobs University in Bremen, Germany, is developing a standard-based, ad hoc, geoscientific data analysis method, scalable to exabyte volume (1 exabyte = 1000 petabytes). The aim is to allow space-time data of unlimited volume to be directly manipulated, analyzed, and combined. The core idea is to define a global, integrated query language for space-time data and their metadata. WCS is also being used in the NASA's GIBS project.

14.4.3 OGC SOS INTERFACE STANDARD

The OGC® SOS provides a standardized interface for managing and retrieving metadata and observations from heterogeneous sensor systems. Sensor systems contribute the largest part of geospatial data used in geospatial systems today. Sensor systems include, for example, in situ sensors (e.g., river gauges), moving sensor platforms (e.g., satellites or autonomous unmanned vehicles), or networks of static sensors (e.g., seismic arrays). Used in conjunction with other OGC standards, the SOS provides a broad range of interoperable capability for discovering, binding to, and interrogating individual sensors, sensor platforms, or networked constellations of sensors in real-time, archived, or simulated environments.

An example of the use of SOS in a big data application is the NOAA Integrated Oceans Observing System (IOOS).† IOOS is "a vital tool for tracking, predicting, managing, and adapting to changes in our ocean, coastal and Great Lakes environment." The OGC SOS is used to provide a standard interface to over 2500 sensor platforms from multiple manufacturers.‡ These sensors are measuring such phenomena as air pressure at sea level, air temperature, currents, sea floor depth below sea

---

* http://www.earthserver.eu/index.php.
† http://www.ioos.noaa.gov//.
‡ http://sdf.ndbc.noaa.gov/sos/.

surface (water level for tsunami stations), seawater electrical conductivity, seawater salinity, seawater temperature, and waves and wind speed and direction. These sensors are collecting millions of observations per day.

Another perhaps more critical use of SOS in big data applications is for tsunami warning. An example is the German Indonesian Tsunami Early-Warning System project (GITEWS). Due to the time criticality related to providing citizen alerts in a very timely manner, completely new technologies and scientific concepts have been developed to reduce early-warning times down to 5–10 min. This includes the integration of near real-time GPS deformation monitoring as well as new modeling techniques and decision supporting procedures. For GITEWS, SOS provides a common and consistent interface to all sensors required for modeling tsunamis.

14.4.4   OGC WPS INTERFACE STANDARD

This document specifies the interface to a WPS. WPS defines a standardized interface that facilitates the publishing of geospatial processes and the discovery of and binding to those processes by clients. Processes include any algorithm, calculation, or model that operates on spatially referenced data. Publishing means making available machine-readable binding information as well as human-readable metadata that allows service discovery and use.

A WPS can be configured to offer any sort of GIS functionality to clients across a network, including access to preprogrammed calculations and/or computation models that operate on spatially referenced data. A WPS may offer calculations as simple as subtracting one set of spatially referenced numbers from another (e.g., determining the difference in influenza cases between two different seasons) or as complicated as a global climate change model. The data required by the WPS can be delivered across a network or available at the server.

WPS provides mechanisms to identify the spatially referenced data required by the calculation, initiate the calculation, and manage the output from the calculation so that the client can access it. This WPS is targeted at processing both vector and raster data. The WPS standard is designed to allow a service provider to expose a web accessible process, such as polygon intersection, in a way that allows clients to input data and execute the process with no specialized knowledge of the underlying physical process interface or API. The WPS interface standardizes the way processes and their inputs/outputs are described, how a client can request the execution of a process, and how the output from a process is handled.

While there are issues with WPS (e.g., profiles are difficult to define), the standard is ideally suited for the big data environment. This is because WPS can be used to wrap any algorithm, calculation, or model. This is why groups have been able to effectively integrate WPS instances with Hadoop. WPS–Hadoop offers a web interface to access the algorithms from external HTTP clients.* This capability was created by the gCube community. gCube is a large software framework designed to abstract over a variety of technologies belonging data, process, and

---

* http://gcube.wiki.gcube-system.org/gcube/index.php/Geospatial_Data_Processing.

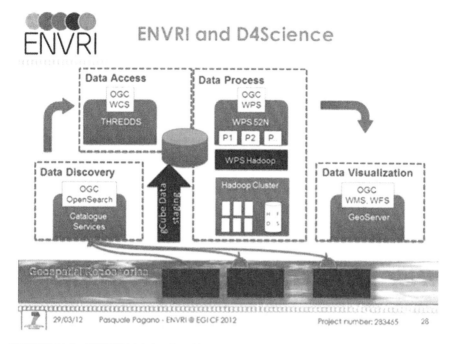

FIGURE 14.2   ENVRI high-level architecture.

resource management on top of grid/cloud enabled middleware.* gCube is a framework dedicated to scientific research.

A related activity that is using OGC standards, specifically, is ENVRI.[†] The ENVRI project, *Common Operations of Environmental Research infrastructures*, is a collaboration in the European Strategy Forum on Research Infrastructures (ESFRI) Environment Cluster to develop common e-science components and services for their facilities. The collaborative effort has to ensure that each infrastructure can fully benefit from the integrated new ICT capabilities beyond the project duration by adopting the ENVRI solutions as part of their ESFRI implementation plans. In addition, the result will strengthen the European contributions to the Global Earth Observation System of Systems (GEOSS). All the nine social benefit areas identified and addressed by GEO-GEOSS will take advantage of such approach. Figure 14.2 shows how WPS is used in the ENVRI framework (Pagano 2012).

WPS is also heavily used in disaster modeling and alerting applications. An example is Distant Early Warning System (DEWS). The DEWS project was funded under the sixth FP of the EU and had the objective to create a new generation of interoperable early-warning systems based on an open sensor platform. This platform integrates OGC Sensor Web Enablement (SWE)-compliant sensor systems for the rapid detection of earthquakes and for the monitoring of sea level, ocean floor events, and ground

---

* http://www.gcube-system.org/.
[†] http://envri.eu/home.

displacements. Data accessed using OGC Web Feature Service instances were then integrated in a simulation system via a WPS instance to identify affected areas.

14.5   KEY ISSUES

Everything we do (and any species on our planet) happens somewhere and *somewhen*. Further, every decision we make potentially has a location (geographic) aspect. Therefore, there is an incredible opportunity in big data analytics to vastly improve decision support for organizations and individuals. There is the potential for predictive modeling that will help scientists and policy makers address the many global issues we are faced with. Predictive modeling has numerous other benefits such as risk reduction, cost reduction, lifesaving, and better provision of services. However, there is also a potential very dark side to geospatially based big data analytics and prediction modeling.

14.5.1   Issue One: Privacy

More and more, the issue of privacy and the use of location-enabled content is in the press. Consider the following two examples of privacy issues that we need to consider when using geospatially enabled big data:

> people's day-to-day movement is usually so predictable that even anonymized location data can be linked to individuals with relative ease if correlated with a piece of outside information. Why? Because our movement patterns give us away.
>
> **Meyer (2013)**

> You can be constantly tracked through your mobile device, even when it is switched off. What's more, those sensors you're pairing with your device make it ridiculously easy to identify you … simply by looking at the data (from the Fitbit) what they can find out is with pretty good accuracy what your gender is, whether you're tall or you're short, whether you're heavy or light, but what's really most intriguing is that you can be 100 percent guaranteed to be identified by simply your gait—how you walk.
>
> **Hunt (2013)**

14.5.2   Issue Two: Provenance

Provenance means the origin, or the source of something, or the history of the ownership, or location of an object. When you access and process big data from a national mapping agency or private sector content providers, you also get information on when the data were collected and compiled, the source of the data, who compiled the data, dates of the most recent updates, and why the data were collected. The same is not true of most Web 2.0 big data. There is no consistent provenance information, if any. Without accurate provenance information, the results generated from a big data predictive analytics process may be suspect.

14.5.3   Issue Three: Data Quality

When individuals or organizations obtain digital map data from a mapping agency or private sector content provider, information on the quality of the data is provided.

These formal mapping organizations can provide scale, measures of accuracy, what coordinate reference system the data are provided in, and so on. They can also provide data quality information on the source data from which the map products were produced. For sensor systems, calibration, accuracy, error propagation, and other metadata are available. The same is not true for much Web 2.0 social media or volunteered geographic data. Sensor-enabled mobile devices, such as your smartphone, do not provide such real-time metadata. Organizations and users need to be able to determine if a specific data source is *fit for purpose*. These application-dependent data use decisions require some level of information on data quality.

Consider two use cases related to data quality and the capture and use of big data. One case is where big data is generated by devices, such as an imaging system, on a satellite. The processes associated with these devices and the quality metadata and provenance are very well known. Highly precise metadata needed to perform radiometric and geometric corrections is available. For such devices, the end user (or application) is able to obtain the necessary measures of data quality to determine if the data are fit for use. At the extreme opposite end are location data collected in crowdsourced real-time situations, such as an emergency event, for which there is little known about the sensors onboard the device used to capture the data. There may be no metadata, no knowledge of the processes associated with the collection of the data, no data quality measures, and so forth. The net result is that the end user or application really has no idea whether the quality of the data is fit for purpose. Further, there may be zero provenance information.

Yet, users and applications will want to fuse data from known devices with strong data quality measures with data with little or no data quality measures. This is exactly one of the issues being addressed in a European project titled COBWEB.

14.6 COBWEB: SENSOR FUSION, BIG DATA, AND DATA QUALITY

COBWEB* is funded under EU's FP 7 in the Environment "Developing community-based environmental systems using innovative and novel earth observations applications" theme. From the project description

> COBWEB is a large project which brings together expertise from 13 partners and 5 countries. The project aims to create a testbed environment which will enable citizens living within UNESCO designated Biosphere Reserves to collect environmental data using mobile devices. Information of use for operational and policy making will be generated by quality controlling the crowd-sourced data and aggregating with Spatial Data Infrastructure (SDI) type reference data from authoritative sources. The infrastructure developed will explore the possibilities of crowd sourcing techniques around the concept of "people as sensors," particularly the use of mobile devices for data collection and geographic information.

COBWEB will leverage the UNESCO World Network of Biosphere Reserves (WNBR). Concentrating initially on the Welsh Dyfi Biosphere Reserve, the project will develop a citizens' observatory framework and then validate the work within

---

* http://cobwebproject.eu/.

the context of the UK National Spatial Data Infrastructure (SDI) and, internationally, within the WNBR, and, specifically, within Greek and German Reserves. The infrastructure developed will exploit technological developments in ubiquitous mobile devices, crowdsourcing of geographic information, and the operationalizing of standards-based SDI such as the UK Location Information Infrastructure. COBWEB will enable citizens living within biosphere reserves to collect environmental information on a range of parameters including species distribution, flooding, and land cover/use.

A main driver will be the opportunity to participate in environmental governance. Data quality issues will be addressed by using networks of *people as sensors* and by analyzing observations and measurements in real-time combination with authoritative models and datasets. The citizen's observatory framework will integrate with evolving INSPIRE-compliant national SDIs and allow the fusion of citizen sourced data with reference data from public authorities in support of policy objectives. To maximize impact, COBWEB will work within the processes of the standards defining organizations. Specifically, the project aims to improve the usability of OGC SWE standards with mobile devices, develop widespread acceptance of the data quality measures, and maximize the commercial appeal of COBWEB outputs. The end result is a toolkit and a set of models that demonstrably work in different European countries and that is accepted as a core information system component of the WNBR. Implementations of COBWEB will act as models for how technology may be used to empower citizens associations in environmental decision making.

A key work area for COBWEB is the relationship between data access and privacy. These are important topics in the development of SDI and ones that are often neglected or misunderstood. Not resolving this issue can also prevent SDI initiatives from meeting their initial purpose. As COBWEB will be linking crowdsourced data with restricted data sources and moving these data to and from mobile technologies while retaining differing levels of security, a scalable access solution is required.

Authentication is the process of establishing that claims made concerning a user who is attempting to use a particular online resource are true. This typically involves confirming a subject's identity with an identity provider, such as a home institution where the user is registered. This is essential in many SDI scenarios, such as when we need to authenticate users to protect privacy or control access to valuable data, and COBWEB will advocate a federated approach to access management. This will allow secure sharing of authentication information across different member organizations' domains and enable single sign on access to multiple web services from different providers.

14.7 SUMMARY

Much of the current data found throughout various communities are already georeferenced. OGC-compliant services can provide properly formatted, spatially aware datasets to web-based analytic applications, facilitating data fusion and the creation of new intelligence products and repeatable analytic algorithms (Hornbeck 2012). There is still much work to be done to learn how to best use OGC standards in big data applications. The OGC is involved in a range of interoperability activities, such as GEOSS and COBWEB, which will allow us to define best practices and enhancements to our standards baseline to better allow the use of OGC standards in big data frameworks.

REFERENCES

Hornbeck, R.L. 2012. Leveraging R and OGC-compliant services. In support of Statistical Analysis, Data Tactics. http://datatactics.blogspot.com/2012/12/leveraging-r-and-ogc-compliant-services.html

Hunt, G. 2013. CIA and Big Data. *GigaOM's Structure: Data Conference*, New York, March 20. http://www.huffingtonpost.com/2013/03/20/cia-gus-hunt-big-data_n_2917842.html

Meyer, D. 2013. http://gigaom.com/2013/03/25/why-the-collision-of-big-data-and-privacy-will-require-a-new-realpolitik. March 25, 2013.

OGC. 2012. A primer for dissemination services for wide area motion imagery. https://portal.opengeospatial.org/files/?artifact_id = 50485

Pagano, P. 2012. Hybrid data infrastructures: Concept, technology and the complex ENVRI RIs use case. http://www.garr.it/a/comunicazione/press-kit/doc_download/1517-presentazione-ppagano

Index

Printed and bound by CPI Group (UK) Ltd, Croydon, CR0 4YY

18/10/2024

01776269-0003